RAINCOAST CHRONICLES 25

RAINCOAST CHRONICLES 25

m̓am̓ała Goes Fishing

Alan Haig-Brown

HARBOUR PUBLISHING

1 2 3 4 5 — 29 28 27 26 25

Harbour Publishing Co. Ltd.
P.O. Box 219, Madeira Park, BC, V0N 2H0
www.harbourpublishing.com

Images courtesy of author unless specified, cover image: *BCP-45* by Bill Maximick
Edited by Noel Hudson
Indexed by Emily LeGrand
Cover design by Dwayne Dobson
Text design by Roger Handling
Printed and bound in Canada

Canada Council for the Arts
Conseil des arts du Canada
British Columbia Arts Council

Supported by the Province of British Columbia

Harbour Publishing acknowledges the support of the Canada Council for the Arts, the Government of Canada, and the Province of British Columbia through the BC Arts Council.

Library and Archives Canada Cataloguing in Publication

Title: Raincoast chronicles 25 : mʾamʾała goes fishing / Alan Haig-Brown.
Other titles: mʾamʾała goes fishing
Names: Haig-Brown, Alan, 1941– author
Series: Raincoast chronicles ; 25.
Description: Series statement: Raincoast chronicles ; 25 | Includes index.
Identifiers: Canadiana (print) 20240510739 | Canadiana (ebook) 20240512316 | ISBN 9781998526185 (softcover) | ISBN 9781998526192 (EPUB)
Subjects: LCSH: Haig-Brown, Alan, 1941– | LCSH: Fisheries—British Columbia—Pacific Coast—History. | LCSH: Traditional fishing—We Wai Kai Nation. | LCSH: Traditional ecological knowledge—We Wai Kai Nation. | CSH: Kwakwaka'wakw—Fishing—We Wai Kai Nation—History.
Classification: LCC SH224.B7 H35 2025 | DDC 338.3/72708997953097111—dc23

CONTENTS

Dedicated to the memory of Mitzi and Herb Assu

Appreciation to Vicki Assu Robbins for permission to use her fine photographs

Thanks to Dana Roberts, Wei Wai Kum language teacher for assistance with Kʷak̓ʷala language

CHAPTER ONE

Introduction

When you hear someone speak of "dancing fingers," most people picture the long, elegant fingers of a flamenco dancer or perhaps those of a concert pianist. But for me the words recall the stubby appendages of a short and stocky captain on a commercial fishing boat. I spent over a decade in my late teens and early twenties with a remarkable skipper, Herb Assu, and the way his fingers could dance a seemingly impossible snarl out of our 220-fathom-long purse seine net stands in my memory as a symbol of all the skills he mastered in the pursuit of his ancestral trade. Those fingers could also deftly handle a knife and a net needle to mend tears in that net with perfectly formed new meshes, as well as feather just the right amount of throttle to close our net around a school of skittish salmon.

Cape Mudge village in the early 1960s, viewed from the *Departure Bay No. 3.*
Vicki Assu Robbins

The fingers were just one part of the man who informed so much of my formative years between 1960 and 1973. Herb was both my skipper and my father-in-law. The two roles were inseparable, and the latter explained the former. At 18, I married Herb's daughter Vicki. Without a high-school diploma, even in those days my employment prospects were limited. I don't recall any discussion. Just as I became a part of the family, I became a part of the crew on the family boat.

The late Herb Assu at home in his world, at the wheel of his family seine boat, the *Departure Bay No. 3*.
Vicki Assu Robbins

Vicki Assu and her younger sister Denise aboard the family seiner about 1961.

CHAPTER TWO

Salmon Seine

In my high school, the kids of fishermen were admired when they returned to class in the fall a month or so late and often driving a fancy car. Suddenly I found myself about to join this elite group, though I would not be returning to school in the fall. But fishing, even then, was not all about fancy cars. Vicki's uncle Buddy Roberts did own a nice new Ford Edsel the day that he set out to cross over from the mainland in the tide-churned and wind-whipped waters off Cape Mudge, at the southern tip of Quadra Island. Buddy also had a wife and several children. As a boy, I had heard of the horrific shipwrecks in these waters, but this was all made real when, in the springtime of 1960, a memorial was held for Buddy. His body was never found, and only ragged bits of his boat ever came ashore.

The memorial began in what had been the one-room school in the village at Cape Mudge. There were several speakers, but there is only one that I remember. Speaking in the Kʷak̓ʷala language, which I do not understand, Chief Billy Assu, then in his nineties, uncoiled his aging frame to stand tall and erect. I understood that he spoke of the powerful tradition of the Liǧʷiłdax̌ʷ people on the surrounding waters. He used his cane to strike the floor for emphasis. I further understood that I was in the presence of powerful people who lived on powerful waters. In a token acknowledgement of the traditional potlatch gift-giving ceremony, at that time much weakened, two or three oranges were distributed to each guest.

Young people were assigned to pick up the dozen or so large wreathes and lead us in a solemn procession through the length of the village to the dock. They then led us down to the float, where we all boarded Harry Assu's classic seiner, the *W No. 8*, and Ivan Dick's beamy *Eva D II*. Tied together, the two boats idled out into the swift-running flood tide in Discovery Passage, a short distance off the village. Once there, the wreaths were dropped from the stern to be carried southward toward the tide rips off Cape Mudge one by silent one. For all of us standing on the decks of those two seiners, it was a solemn, moving ceremony. It was a sea-silence that I would come to know more of, and one that I would never forget in all my time at sea.

That first summer, we fished salmon with the 77-foot *San Jose*. I learned that the boat had belonged to Herb's father, Dan, until he drowned when he drove his car off a wharf. Her hull was newly painted green to signify that she now belonged to the Nelson Brothers fishing company. But the company, owned by two Norwegian immigrants, saw mutual advantage in having the Assu family run the boat. Herb also owned the smaller 57-foot *Departure Bay No. 3* (*DB No. 3*), which was skippered that year by Art Chickite, who, like Herb and his family, came from Cape Mudge village of the We Wai Kai people, members of the Liğʷiłdax̌ʷ nation.

San Jose, the boat Alan crewed on during his first year, here taking on fresh water while tied to a bluff at Knox Bay. The 77-foot *San Jose* was built by Jirokishi Arimoto in 1928.

The *DB No. 3* was built by Jimburo Sawada on Nanaimo's Newcastle Island in 1928 for the Japanese Canadian twin-seine herring fishery. This was a technique that these fishermen had brought from their homes in Wakayama Prefecture. Many years later, my Japanese-speaking wife, Ananya, and I would discover a pair of such boats on a beach

Above: The *Departure Bay No. 3* was built in 1928 by Jimburo Sawada of Nanaimo to seine herring in Georgia Strait before being acquired by Herb Assu, who specialized in the Johnstone Strait salmon fishery. In this 1963 photo, the fenders have all been moved to the starboard side so the net could be worked on the port side. The *DB No. 3* was lost off the mouth of the Skeena about 1980.

Left: Believe it or not, these are two seine boats photographed by Alan in Wakayama Prefecture, Japan. "Twin seining," using two seiners on one net, was a practice imported to BC by immigrants from Wakayama.

near the Wakayama town of Mio, from which many had immigrated to Steveston. The original owner of the *DB No. 3*, Ryutaro Kita, also salted the herring for export to Japan. Like so many other Japanese Canadian vessels, the *DB No. 3* was confiscated by the Canadian government following the Japanese attack on Pearl Harbor in 1941. She was sold by the Canadian government into a fishing company fleet, and Herb bought it from there. I was told by older crewmen that *DB No. 3* was a seized Japanese Canadian boat, but it would be decades before I was able to trace the full story.

In high school, the Indigenous guys, with their good looks and with money from summers spent commercial fishing, were the leaders of the "cool" group. I knew and occasionally hung out with them but didn't have that level of cool. I soon saw more of that group, as several were members of the Assu/Roberts family that I had married into. As the child of immigrants—my mother from the US and my father from the UK—my cousins were distant ideas. Now, through marriage, I became part of an extended family with dozens of relatives, old and young.

The *San Jose* was very much a family boat. Herb was the captain. His wife, Leoda "Mitzi" Assu, cooked three big hot meals daily with the diesel-fired stove in our compact galley. Herb's brother-in-law, Jimmy Mitchel, was the deck boss. He was in his forties, moody, and could be impatient with greenhorns like me. He also possessed a wealth of knowledge, which he doled out when he saw fit. It was Jimmy who taught me to splice the heavy four-strand manila purse lines that we used.

Over the decades, I have added to Jimmy's teachings. Company-owned boats were painted distinctive colours: Canfisco used black hulls and reddish cabins, BC Packers had black hulls and buff cabins, while Nelsons had green hulls and white cabins. When a fisherman bought a boat, he usually painted it all white. One day I observed to Jimmy that a particular white-coloured boat must be privately owned. "Just 'cause it's painted white don't mean he owns it," Jimmy said bitterly. It was an important intro to corporate debt control of so-called "independent fishermen."

I was to become more conscious of what I would call "corporate debt bondage" in the fishing industry. The fisherman's place was precarious. We had organizations like the United Fishermen and Allied Workers' Union (UFAWU), the Fishing Vessel Owners' Association and the Native Brotherhood to help. The power of the fishing companies over the workers was great because we had to catch migrating fish and needed an immediate market that only large processors could provide. In addition to the influence of federal regulators and corporate lobbyists, there were so many other variables that vessel owners had to take care to maintain in their company relationships. This was not only for the "free turkey at Christmas," but also the need to source the thousands of dollars in credit required to maintain and update their boats and gear.

One time, when Jimmy Mitchel was still the deck boss, we were travelling south past Kuper (now Penelakut) Island. There was a long pier reaching out from an imposing building on the shore. I asked Jimmy what it was. "That is the residential school that I went to," he said with tense anger in his voice. "One time we tied up the boat there, and I walked up and punched that fuckin' brother in the face." I knew about the punishment that children had received for speaking their language at these places, but it would be years before I learned of the other horrors. I didn't ask Jimmy more and he didn't offer, but his short account of the incident has stayed with me for decades.

That first season, we had one other white guy, Graham, who was not related to the family but who had been brought on to work with me in the skiff. Two of my young, barely teenage brothers-in-law were half-share crew (meaning they were only paid half as much as regular crewmen) and, when not fighting with each other, helped on deck. A third brother, four-year-old Darryl, rounded out the usual crew.

Mitzi was the cook that summer and for the dozen summers that I spent on the family boat. The second and other years we were on the family's smaller *DB No. 3*. I would guess that the galley was about 8 by 12 feet. There was a table that could seat five adults with room for one small child. Once the meal was served, Mitzi would go out on deck or up to the wheelhouse if it was raining. Space was limited.

Above: When this photo was taken in 1990, the Area 13 fleet based in Campbell River and Quadra Island included most of the same boats as in 1970, when Alan fished with it. Current trends will see the last of these wooden classics replaced by half the number of aluminum, steel and fibreglass boats.

Left: Alan piling corks on the *DB No. 3*. As the net was pulled in using the power block, it had to be neatly stowed on deck with the cork line on one side and the lead line on the other. Piling corks was the preferred position, as you could stay dry. *Vicki Assu Robbins*

Alan resting after an early morning set on the *DB No. 3* in Johnstone Strait. *Vicki Assu Robbins*

And, oh my god, those meals that she made. After the first set of the season, Herb would come down on the deck and select a beautiful, bright humpback salmon female. We would carefully dress her and keep the eggs in a bowl. These we gave to Mitzi, who would make humpback stew, an elegantly simple meal of lightly boiled cubes of fish, potatoes and onion. This was served with generous dollops of *ƛ̓ina* (eulachon grease). It was like the First Salmon ceremonies that I would learn about later in university. Now I treasure the memories of those meals that Mitzi made as the most glorious and loving ceremonies I have ever shared.

In summer we usually started work around 4:00 a.m. in July and 5:00 a.m. in August. Breakfast would be about 8:00—stacks of pancakes, fried eggs, bacon and, on occasion, fresh buns. Lunch typically included soup and a main dish, from spaghetti to chops. A few hours later, it would be supper, with a packed table of meat and potatoes. Despite the abundant food, the hard work kept most of us from gaining weight.

But it was not all work. In the 1960s, fishing was in some ways a gentler profession. Herb liked to set certain places, like the Slide or farther down at Ripple Point. But our favourite place was just below Mitzi's Bay. He would anchor out just off a fine gravel beach, in the bay named for his wife, to wait for the tide to turn. We young ones would take the skiff to try and jig a cod, or beat our way through the salal bushes that lined the beach to climb the hill and pick huckleberries. We would bring them back to the boat, and Mitzi would make huckleberry pie. Other boat cooks marvelled at the baking that she did in Mitzi's Bay. Cake, pie, bread, all came out of the tiny oven on the little diesel stove in the corner of that crowded galley. Summer westerlies blew in the window on one side and out the Dutch door on the other side. Temperatures changed constantly, but the food was consistently amazing.

By August, the big Fraser River sockeye were coming, and Mitzi would trim two or three in the classic manner. Cutting in from one side of the backbone then turning the fish over to cut in from the other, she removed the bright-red sockeye flesh in one large piece. Herb would mount these with a lattice-work of cedar strips, then slide them between larger pieces of wood to form a roasting frame.

We took these slabs back to the gravel beach and built a fire, which we let burn until we had good coals, then the sticks would be set around the fire. We watched as the oil in the fish began to boil and drip slowly down onto the gravel. When it was done to perfection, we loaded it in the skiff and took it back to the boat for a feed. There was usually enough of this *ƛubəkʷ* to last for a day or so. Thank you, Mitzi. That will always be my finest meal.

On my second season of salmon seining, Herb was back in charge of his own boat, the *Departure Bay No. 3*. It was designed to catch salmon on their migration to their natal streams, but I came to understand that the boat's role was much larger than simply a vessel for catching salmon. The ancestral Assu family had done that for thousands of years. The *DB No. 3* was also a family boat among a larger family of boats. Herb's brother Ronnie was married to Mitzi's sister Lila. At night, after delivering the day's catch to the packer, we would often anchor in the shelter of Knox Bay with Ronnie and Lila's boat, the *W No. 4*. During the day, while waiting for the tide, we might lay alongside boats run by Mitzi's brothers, Tony, Aubrey or Gerry, or Mitzi's brother-in-law Charlie, or other family boats.

A typical July day started with the sound of Herb coming down into the engine room, just aft of where the crew slept in four or five bunks in the fo'c'sle. It was his way to check his engine before starting it. As family head, he didn't designate one of us to be engineer but took that duty on himself. Hearing him in the engine room opening the squeaky valve for the gas-powered starting engine was my signal to jump out of my upper bunk. By the time the big old Cat diesel was throbbing, I was dressed and wide awake. I would then make my way aft through the engine room, up the ladder, and through a trap door into the galley before going out to the bow. There, I took up an iron bar and guided the anchor cable and chain onto the drum of the anchor winch as Herb activated the hydraulic control to haul up the anchor. Few words were exchanged, though at some point over the years, after reading a book about seine fishing in Chile with this as a title, I began saying a ritual "Good morning, Captain" in a playfully formal manner. On fishing boats, "skipper" is the more usual term.

Once the anchor was up, at 4:30 on a July morning, dawn was hinting at the sunrise that would soon break over the mountains and islands to the east. In August and September, the shortening days gave us a little more sleep. Herb, square and squat, had a body designed for a maritime life. I have seined salmon with Herb from Whale Channel and Milbanke Sound in the north for pinks to Satellite Channel in the south for chum. But his preferred waters were a 12-mile stretch of Johnstone Strait, along the Vancouver Island shore from Camp Point in the west to Ripple Point in the east.

There are no typical days on a fishing boat, but there were good days, when the wind and rain didn't drive too hard, the tides were regular, and we caught our share of beautiful Fraser River–bound 7-pound sockeye. Such a day started with the anchor up and

the boat drifting just off the Vancouver Island shore by 5:00. Someone made a pot of seine-boat coffee by boiling a goodly amount of coffee in a small amount of water in the big, enameled coffee pot. Once it had boiled enough to spread the coffee scent through the little galley, boiling water from a kettle was added to make a full pot. A dash of cold water might be added to settle the grounds. If weather forced us to anchor across the strait in Knox Bay, on West Thurlow Island, this could be done while the boat travelled to a favourite site known as the Slide, just below Camp Point on the Vancouver Island side. If there was time, the crew could down a quick cup of coffee before the skipper told us to stand by for the day's first set.

In the early 1960s, all things were governed by the tides. There are four slacks per day, two southbound floods and two northbound ebbs. The salmon travel on their migration toward the Fraser or other smaller rivers along the south coast, often following the shoreline. To catch them, we anchored one end of our net close to shore by fastening a beach line around a stump or a tree or even a rock. In special places, boat crews had installed "pegs," heavy pieces of drift log anchored in a crevice and braced with big rocks.

On a typical day, the low- or high-water slack occurred around dawn. Herb set the boat just offshore and, with one or more of the crew, began scanning the water's surface for the dorsal fins of salmon rolling like miniature orcas just out of the water. This finning is most common with pink or humpback salmon. Sockeye like to launch themselves out of the water before sliding back in on their sides. We call them "sliders" or, if they come higher, as the pinks do, the crew will yell out, "Jumper!"

Herb learned his techniques from his father in the 1940s and '50s and was not quick to adopt new ways. He took pride in his ability to read the tide and the movement of the fish. He also had a deep love for the home waters that his father, Dan, and grandfather

Opposite top: The *W No. 4* running. The boat belonged to Mitzi's sister Lila and Herb's brother Ronnie Assu. It was brought up from the US about 1914 and was always kept in immaculate condition by Lila.

Opposite left: Lila Assu in the *W No. 4* skiff. Lila was a fan of Jackie Kennedy in the 1960s, and she was a very stylish dresser, even on the boat. *Vicki Assu Robbins*

Opposite right: An early selfie of Alan (left) and two mates in the fo'c'sle on the *DB No. 3*.

Opposite below right: Built in 1964 by Benson Bros. Shipbuilding in Vancouver and owned by the late Tony Roberts, the *Western King* is always maintained with pride.

Rowing the skiff for tie-up man Jack Quatell at West Thurlow Island, probably late fall in 1960, as Herb fished that side for chum salmon. Alan painted his oar tips and hard hat to match the trim on the boat. *Vicki Assu Robbins*

Chief Billy had seined since at least as far back as 1920. I was once told that Chief Billy, fishing one of the early cannery boats from Quathiaski Cove, on Quadra Island, would complain to his crew, "I can see more with my glass eye than you guys can see with two good eyes!"

I don't recall Herb yelling, "Jumper!" He was more likely to point a stubby forefinger at the place where a salmon had shown. If the tide was right, and I was standing up top on the flying bridge with him, he would say, "OK, we'll try it."

That was my signal to scramble down the ladder on the side of the cabin and walk quickly across the deck, climb up and over the web pile, and then down into the heavily built, beamy 18-foot skiff that was held tight to the stern with its bow drawn up by the painter (tie-up rope). Some years I was the "skiff man," so I would scramble to the back of the skiff then stand facing forward with the 9-foot oars at the ready. If I was "tie-up man," I would stand in the bow of the skiff ready to catch the skiff's painter when the deck boss, following Herb's wave to "Let 'er go!" would release the skiff's line from the deck cleat.

Herb would run the boat full speed toward the steep rock mountain-base that forms the "beach." It is deep along the shore of Vancouver Island here. At the last moment, he would spin the wheel on the flying bridge to swing the stern as near to the beach as

Letting the skiff go as the *Departure Bay No. 3* starts a set in the early 1960s. Timing was everything, as a table seiner began its set by releasing the heavy skiff to pull the net into the water. *Vicki Assu Robbins*

Far left: Tie-up man Tarzan Scow and skiff-man Jay Grabher tying on the beach for a set by the *Adriatic Star* in 2010.

Left: The tie-up man prepares to release the slip knot on a tight beach line.

Far left: The tie-up man secures the beach line to a tree before the full strain of the towing seine boat tightens the line.

Left: After letting the end go from the beach, they return the end of the net to the seiner. The aluminum skiff and the outboard have changed the job, but the sea is unchanging.

Opposite top left: Drying up the net in 1963 in preparation for brailing. Left to right: Joan Assu on cabin top, Herb Assu, Alan Haig-Brown and Harvey Assu.

Opposite top right: Piling corks on the DB No. 3. The crew took extra care on the last set of the week to make a "weekend pile" that would look good in port. *Vicki Assu Robbins*

Opposite bottom: Herb and Alan pursing up. Herb didn't normally purse, but he didn't trust the new deckhand who he just hired off the dock in Campbell River. Note: Alan is pursing with a twisted style line, while Herb is pursing one of the new braided lines that didn't have to be coiled. *Vicki Assu Robbins*

practical. In rapid succession the deck boss would release the skiff—a heavy line from the end of the net was tied to the side of the skiff with a slip knot. Upon release, the line would turn the skiff broadside to the boat's stern and toward the beach. The weight of the skiff would pull the net off the stern and, at the same time, the skiff man would dig in the right oar to help turn the skiff. The tie-up man retrieved the skiff's painter and released the slip knot once there was enough net in the water to drag the rest of the net off the boat. If it all worked well, the skiff man gave a few strokes of those big oars to ride the wake of the boat and quickly get the skiff to the shore. When the bow of the skiff touched the shore, the tie-up man leaped to the rock face and scrambled up to the "peg" to make several wraps of the beach line. Speed is important to assure that the end of the net is kept as close as possible to the beach. Every tie-up spot is different, and the two skiff men must be well coordinated.

Doing all of this with heavy manila lines and fast enough to keep the bunt end of the net (the part where the fish would ultimately be concentrated) close to shore was always an adrenaline rush. Often kelp beds had to be crossed, or an especially steep cliff had to be climbed, to add to the excitement. Once the seiner had run out all 220 fathoms of net, it would start to tow against the gentle beginnings of the flood tide, and the beach line would tighten, lifting the end of the net out of water. The beach line tightened until drops of water began to be wrung from it.

The skiff man rowed out to the end of the net, if the strain hadn't lifted it into the air, and began plunging. This involved throwing an 18-inch length of 2-inch pipe into the water and retrieving it, making bubbles to keep the fish from coming around the beach end of the net.

Regulations and practicality required that the seiner tow the open net, holding it at maximum width, for only 20 minutes before circling up against the current and back to the beach to enclose whatever fish it had intercepted. The adrenaline would peak again as Herb waved to let the end go from the beach. The beach man had to quickly remove the wraps from the peg (in later years this entailed releasing a complex slip knot) then scramble down the rocks to jump in the bow of the skiff. While the skiff man rowed, the beach man pulled in the extra line, uncoupled a figure-8 shackle and then passed the shorter length of line up to the man on the bow of the seiner.

We two would then climb up over the bow of the seiner and go quickly to our assigned jobs on the main deck as the winch pursed up the bottom of the seine. I took pride in my agility at leaping around the skiff and the boat, but I also admired Herb's solid presence on deck. He had the physique of one whose ancestors had been canoe men. Short legs, heavy body and powerful arms. Once, in Ocean Falls, some young guys were showing off with those exercise things that have two handles connected by five or six springs. They were stretching it across their chests. Herb tried and, holding it at arm's length, pulled it to its full extent. It took those kinds of arms to pull a canoe paddle all day or to do the climbing of ladders and pulling of lines required of a seine fisherman.

Herb generally came down on deck when pursing up the seine. Like most skippers, he took control of the plunger, a long aluminum pole with a metal cup mounted on the end. He would drive this into the water beside the boat where the net was not yet

pursed. The cup made a satisfying *kurpup!* with each stroke of his powerful arms. This was to scare fish away from the opening in the net where they might escape. All the while he would be watching for bubbles or movement along the cork line to signify that we had fish.

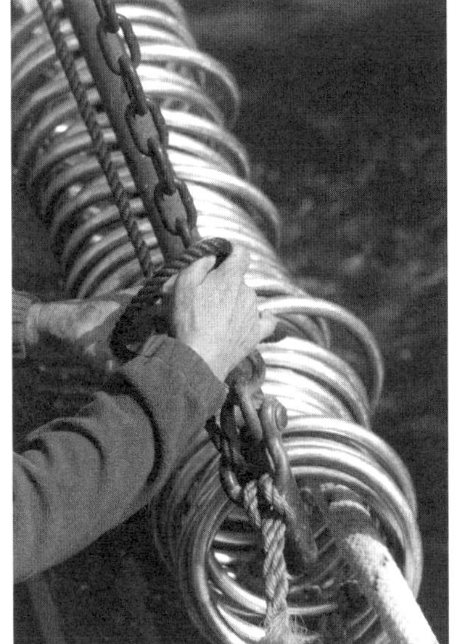

The purse rings are spaced along the bottom edge of the seine so it can be "pursed" or closed with the purse line, trapping the catch.

Once the net was formed into a circle like a fence around the fish, the next move was to close up the bottom by pursing, a process similar to tightening the string of a drawstring bag. The bottom edge of the net was fitted with brass rings with a line through them. This "purse line" was pulled tight using the deck winch. Once the purse line was all in, the closed-up bottom of the net, with its lead line and brass rings, would be lifted onto the deck. Whatever fish had been encircled were now trapped inside the 220-fathom net with its floating cork line, the bulk of which was still in the water.

Throughout most of the pursing procedure there would be web near the propeller, so the engine could not be engaged without tangling the net, and thus the boat's movement was largely at the mercy of the tidal current. If necessary, the boat could be shifted a little by pulling on the net and piling the corks, with attached web, onto the bow or stern. With the bottom of the net drawn tight by the purse line and lifted onto the deck, we had only to haul the 220 fathoms of net back aboard using the hydraulic power block until all the fish that had been encircled were concentrated into a small space in the bunt end of the net. If the catch was not too large, the bunt could be rolled aboard with all the fish in one move, but a big catch would have to be brailed aboard 100 or 200 at a time using the big power-assisted dip net or brailer. I didn't fish in the times before power blocks, but those who did told of brutal work. Those earlier boats were called table seiners because

After a set, the net would be hauled aboard until the catch was pooled in a reinforced pocket called the bunt. The fish would then be scooped out with the brailer, a large power-assisted dip net with a closure in the bottom.
Vicki Assu Robbins

Above: The skiff returns the net's beach end to the seiner to complete the circle.

Left: The hydraulic power block, invented by Mario Puretic in Seattle, was introduced in the decade before Alan began fishing. It made life easier, but wind and tide could still twist a boat into its net. Alan would have pulled corks into the skiff before going to pile them as the power block recovered the net. *Vicki Assu Robbins*

they had a large turntable on deck with a roller along one side. This table would be rotated to face the net, which the crew would pull aboard by hand. It took seven strong men. The heaviest job we had was piling the net on deck as it came through the block.

Other than a large chunk in the meat cooler on top of the boat's cabin, we carried no ice for preserving fish on board. The day's catch had to be delivered to a company packer each night. For a greenhorn, this posed some challenges in identifying the different species while quickly pitching them from the hold onto the deck to be pitched again into the packer's scale.

I knew how to identify humpback (pinks) and springs (chinook), as they spawned in the river near our house when I was growing up. My father caught an occasional steelhead with his fly rod. And dogs (chum) came into the river in the fall. Coho were common in the small creeks around my childhood haunts and were also caught in great numbers by visiting sport-fishermen, who freely gave them away. I knew that there were a very few "creek sockeye" in our Campbell River, but the species needs an accessible lake upstream to rear and thrive. Sockeye, which normally feed on plankton, were not known to take the hook in those days. The first time that I saw a good catch of these firm, silver-bright 7-pound fish on the boat, I was amazed. When I saw their brilliant red flesh, I understood the reverence with which Herb and all the crew regarded them. Soon, like the rest of the crew, I was taking one or two home on the weekends.

Pinks, coho, springs, chum and sockeye all fetched different prices and had to be announced as they were pitched aboard the collector to create a correct tally. The small pinks were easy, though the less common coho could be confused if the same size. It was fun to get good at it, and even more fun when the guys on the packer would debate an individual fish.

As for killing so many fish, I mostly didn't have any qualms. The delicately coloured and infrequent steelhead caused a little reflection, as they were generally entered on the tally as coho. That seemed indecent for such a special fish. The hard ones were the big springs of 40 pounds or more. They came as individuals rather than in schools. For a short time after death, they kept their golden iridescence and seemed to stare at me with a big, baleful eye that made me feel queasy inside, something that I would never have admitted to the crew.

CHAPTER THREE

Up North for Pink Salmon

"You have it easy," I was told by the older crew who had pulled nets by hand before the advent of hydraulic power blocks. Over my years on the boat, hauling the net back in got still easier. First, once Herb shifted away from the old cotton web, the lighter nylon mesh reduced the sheer volume and weight coming through the power block. Then the big hydraulic drum on the stern removed the need to pile the net. But my memory usually takes me back to the early years. I suppose the same would have been true for Herb. He was not a storyteller and said little about how it was in his youth. But in his calm and cautious manner, I came to understand the depth of his knowledge of tides and fish.

That knowledge served him well when we travelled to other fishing grounds. July was often a time of scratch fishing for pink salmon in his favourite waters. The company that we fished for often wanted us to go and join a northern migration of boats to fish the plentiful but low-value pink salmon in northern waters around Caamaño Sound or on the central coast near the cannery at Namu. Herb was a fisherman, not a businessman, so the company would lend him money to get the boat and gear ready for each summer's salmon season. Debt gave them some control over where he fished. I remember the company dispatch boat coming alongside and conversing with our skipper.

The dispatch boat would leave, and Herb would say, "OK, boys, pull the skiff up." He would climb the ladder to the flying bridge, turn the big brass wheel to put the boat in gear and then the smaller wheel that was the throttle. With the skiff securely stowed on top of the net, the boat would head "up north," which on the BC coast is actually more westerly.

When a crew gets it right and develops as a team, the work can be a lot of fun. Rain or shine, lots of fish or few, we enjoyed the rhythm of working smoothly together. One summer it went wrong from the start. We were late getting away, and Herb hired two white guys off the dock. Magoo had spent a season on a seiner, but his sidekick had not done even that. Unpacking his duffle bag, Magoo took out a pistol and stashed it under his mattress. I thought that he was just showing off but learned later that he was also dealing heroin.

Herb assigned them to the skiff, which meant I got to work on deck for a change. The deck was easier than being in the skiff, but I quickly saw that Magoo was giving his greenhorn friend the harder jobs when the opportunity afforded. We went north to fish Milbanke Sound and Whale Channel. It was a typical year, with good catches of small 2-pound pink salmon. It is the kind of fishing that requires some work but that can give you a good start before fishing sockeye in Johnstone Strait.

We fished all day, sometimes tying on the beach but often making round sets in open water, which is not so dependent on the tide. We might catch 5,000 or more of these small, much-maligned fish. A long day of fishing, then waiting for our turn to deliver to the packer could make for a tired body. Climbing down into the hatch to stand up to your crotch in slimy humpback was not something to cherish.

Built in 1944 for the Canadian Air Force, *Hesquiat* and her sister *Kimsquit* were sold to a fishing company and worked as fish packers for decades until being sold off by Canfisco sometime after 2020.

In the 1960s and '70s we used a fish peugh (or pugh), a tool like a single-tined pitchfork with which you could spear and toss a fish. With the higher-valued 7-pound Fraser River sockeye, the packer crew insisted that we peugh the fish in the head rather than the body, where we would damage the meat. With large catches of humps, they typically lowered a cargo brailer into the hatch and we speared two or three at a time into the brailer, to be lifted and dumped onto the packer's scale and weighed. Speed was more important than quality. By the 1980s, quality control required that all fish had to be transferred by hand, what we called "hand bombing." When off-loading at the cannery, the fish were transferred by Transvac pumps that sucked them up in water without any damage or human contact.

Meantime, back at the packer in Whale Channel, I noticed that when we delivered late at night, Magoo always managed to put himself up on the bow, where he would tie up the bow rope then linger just long enough that his poor partner and I back on deck had no excuse but to jump into the hatch and do the dirty work. He never offered to spell us off. We would have to spend time after we came out washing each other off with the deck hose. On at least one night I neglected to put rubbers around my slicker pants and ended the night with boots full of fish slime. These had to be washed and hopefully dried by the next morning.

One evening I went up on the bow and stood fast with the bowline, then hid around the cabin until Magoo had no choice but to jump into the hold. I found that version of office politics so stressful that the next night I jumped into the hatch with renewed vigour. Herb said nothing, but on our first trip back to Campbell River he fired them both. His quiet way.

CHAPTER FOUR

Navigating the Coast

We had no autopilot to maintain a course. When travelling, the norm is for each crew member to take a two-hour turn at the wheel. At night this changed to two crew on watch for four hours at a time. In the early days, before we had radar, we travelled from point to point by day and from light to light by night. The whole of the BC coast has navigational lights to help boats steer courses from point to point. These were all listed in Capt. Lillie's coast guide, with details on the distinctive patterns of their flashes and the distance to safely pass off from them. In winter, when travel was more frequent and nights were long, the Capt. Lillie was kept ready on the compass table.

Once, in summer, one of the sons and I were on the wheel for some time before and after the dawn. When Herb got up from a four-hour sleep, he took a look out the wheelhouse window. Waves were breaking over a rock far off to starboard. "Oh," he said, "you came this way. I haven't been here for years."

The wheelhouse interior of the *DB No. 3*. The large brass wheel is the clutch and the small one is the throttle. Chain steering required a heavy pull on the big spoke wheel.
Vicki Assu Robbins

He could recognize where we were with just one glance. Another time, I was steering up the Fraser River past Steveston. When headed up a river, you must keep the green buoys to port and the red buoys to starboard. It was partial darkness, and I was worried, but it is always scary to wake the skipper. No sooner had I begun to worry than Herb was at my side. "I'll take it," he said, reaching for the wheel. A couple of days later we were in a bar in New Westminster. I praised his timing at taking over just when I felt unsure. He maintained that he didn't know what I was talking about. He was not fond of verbalizing the skills that made him a great skipper.

Herb taught by example and by letting me try things on my own timing. This was very different to the detailed verbal direction that I had been raised with. I recall another time when I was on the wheel up top as we entered Seymour Narrows headed south with a strong flood tide pushing us to nearly twice our normal speed. There were tide rips, huge whirlpools and, between them, to port and starboard, huge glassy boils welled up from the depths. I nervously tried to steer my way through them all and to keep the boat from rolling too much. Finally, I heard a dish crash in the galley. There was a gentle tug on the wheel, signalling that Herb would take over from the wheelhouse. Relieved of control and my nervousness, I was able to relax and watch as he masterfully navigated the boat through this notably treacherous piece of water. I marked his movements and stored them for future use. Not a word was exchanged.

Navigation on the BC coast is largely about local knowledge. There is what we used to refer to as the steamboat channel, a deep-water route up the middle of broader passages used by large ships for the full length of the coast. Fishermen in boats that seldom

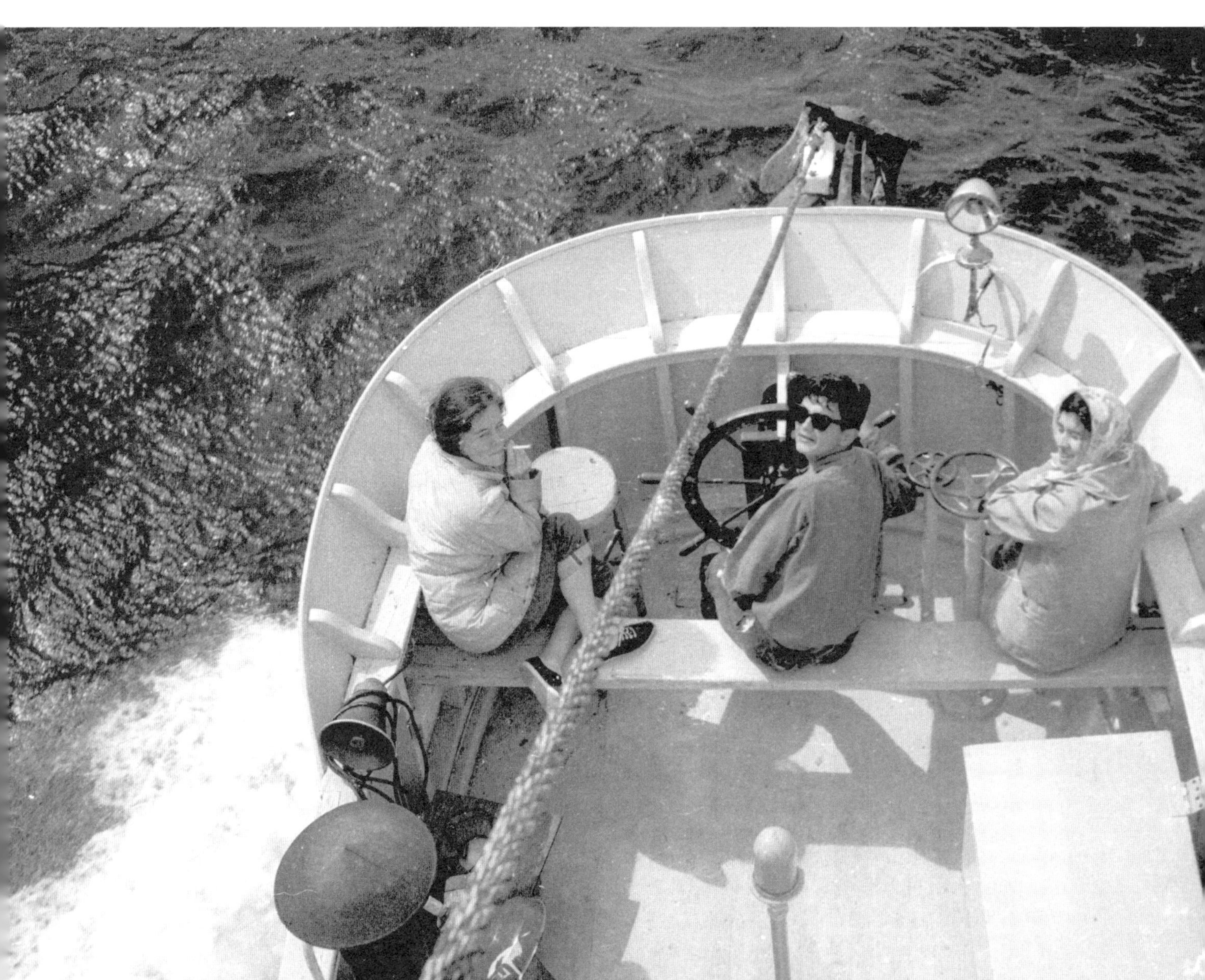

Three Assu siblings, (left to right) Vicki, Wayne and Joan, enjoying travel on a sunny day.

exceeded a 10-knot speed looked for short-cuts. It took me some time to realize that, while Herb was confident in his knowledge of many of these, there were places that he didn't go. I think that he had learned directly from other fishermen, like his father. For instance, Meyers Pass is a narrow channel behind Klemtu and Swindle Island. It is a good short-cut to Laredo Sound. You enter going due south, then, halfway through, you turn 90 degrees and pass over a shallow spot where you can see the bottom. We went through there several times, but we never went down the north arm of the Fraser River. I don't know why. Gunboat Pass, near Bella Bella, with a huge boulder just under the surface, was not a problem, though he did go slow and ask me to look out from the bow as we drew near.

It wasn't until I spent time on small coastal freighters that I learned about charts and logbooks. On those boats, we stood six-hour watches, a deckhand and the captain opposite a deckhand and the mate. The appropriate large-scale chart was always out on the chart table. As we passed each significant point or feature, it was entered into the logbook, as was each docking and watch change. I did once see a chart on the seiner, but it was stored under the skipper's mattress in the wheelhouse. It was so old that the shoreline of many bays and inlets was marked by a dotted line to signify "uncharted." Herb carried his charts in his head. Years later, I purchased a copy of the very detailed *British Columbia Coast Pilot, Volume 1* for the south coast. It gave detailed navigational instructions for each inlet and passage. I very much doubt that Herb had ever seen a copy or, had he, I doubt that he would have been impressed. The printed word was not his way in the world.

I knew that Herb had run away from the Coqualeetza Indian Residential School in the Fraser Valley after attending only a couple of years. But I also knew that he carried a good part of the *British Columbia Coast Pilot*, that bible of coastal navigation, in his blood. He carried a feel and knowledge of tides and currents that no book could replicate. I swear that he could feel more through the soles of his feet, set firmly on the top deck of the boat, than all of today's modern electronic navigational aids could match. I don't recall ever seeing Herb read anything except a tattered tide book, and that was only when he had been off the boat for a time and had to get reoriented.

Once back on the boat, he knew closely when the tide would turn at any given place. The tide book gives good estimates for specific spots on the coast, but to adjust for a spot 20 miles away requires local knowledge. The closest that Herb came to verbally instructing me in the art of tides was to ask, "What do you think, is it ready?"

I felt pride that he would ask me if the tide was ready to set. I scanned the heads of the bull kelp along the shore. They were all just starting to stream south. "Yeah, it looks good," I said. But then Herb pointed at a faint line, indicating a tide rip just offshore where the net would reach. "No, I don't think quite yet," he commented without any judgement of my imperfect knowledge.

Coming, as I did, from a highly verbal family and culture, I didn't learn well under this tutelage. Thirty years later, I went out fishing several times with Darryl Assu, one of Herb's younger sons. We fished the same waters and the same tides. We now had a modern drum on a different seine boat, but I watched Darryl gently move the big boat with a touch of wheel and clutch as it drifted close to shore. I saw how he observed the tide and, finally, how he gave the order to his crew to "Stand by!" I was no longer a deckhand but a fishing journalist taking photos and notes for a magazine.

Now, I had the time and freedom to observe the whole operation. It was uncanny. The years had peeled away. Instead of being younger than my skipper, I was now a good bit older. Herb had died years before, but his teaching, like the tides, continued. I also have come to understand that this knowledge of the tides has not just evolved since the introduction of power boats and purse seine technology around 1912, it is set deep in the inner understandings of the Indigenous people.

When I was first married, before I went fishing, I was finding my way in Cape Mudge, and an elder named Jimmy Hovell gave me a job helping to take down his family's old house so that they could build a fine new one on the same foundation. He was such a kind and gentle man. I think that he was just giving me a chance to get to know others in the village and the routine of those times. The first summer I got out on Herb's boat, Jimmy skippered, as he had for years, a smaller 50-year-old Nelson Brothers boat called the *Gradac*. During the season he had heart trouble and had to stay home one or two weeks. The next summer, the company didn't give him a boat to run. I would see him shuffle down to the dock at Cape Mudge to watch the boats go out. I was 19 and in peak health, but I could feel his pain. As I watched I remembered his sorrow when, a few years before, he had shown me a worn little picture of a boat that he had owned ablaze at the dock. As much as a 19-year-old can, I felt this good man's sorrow at having to stay home from the fishing grounds. I understood how his very identity was in the tide and the fish. He died that winter.

In the years that I spent with the Assu family, I had the opportunity to know some elderly knowledge keepers whose memories reached back before 1900. Besides Jimmy Hovell, there was Chief Billy Assu and Jimmy's wife, Louise Hovell, who we knew only as Autie. Autie was the grandmother of Herb's wife, Mitzi. She told me that she had been born at a cannery on Canoe Pass, in the Fraser River estuary, where her family had gone for summer work. It was, she said, the year of the Great Vancouver Fire, 1886. She also told of travelling by dugout from her family home at Salmon River, just to the north of Herb's favourite fishing spots. Tides of 2 knots and more are a constant in Johnstone Strait. To paddle a big dugout south from Salmon River, the people would have used the same back eddies and currents that Herb used to set his net. Some memories are inscribed as solidly in the mind as the gouges that the glaciers left on the rock face of the beaches. These are memories that exist beyond generations and time.

I recall one November when we were travelling south to fish dog salmon. When a thick fog came with darkness, Herb decided to anchor out and wait for the fog to lift at daylight. The morning broke bright and clear. We were anchored in some small, rocky islands just north of Departure Bay, and it was an easy run into Nanaimo. Years later, Autie sang a song that she had learned while travelling these same waters by dugout. When the fog set in, she recalled, the people in the canoe kept paddling while singing in Kʷak̓ʷala, "Where are you, Victoria?" She sang the song for me, keeping time with the back of a hairbrush on the kitchen table and, in her beautiful high voice, pronouncing "Victoria" as "Mitoria."

Above: Left to right and back to front: Mitzi Assu, Judy Assu, Jimmy Hovell, Mrs. Roberts, Mrs. Autie Hovell, circa 1966. Mrs. Roberts was Vicki's grandmother, and Autie, who was from Salmon River, was Vicki's great-grandmother. *Vicki Assu Robbins*

Left: Darryl Assu, about five years old, on the wheel of the DB *No. 3*, getting an early start in the world of boats and tides. *Vicki Assu Robbins*

CHAPTER FIVE

Fishing Heritage

In 1983, a decade after my last season of seining, I began to research the history of commercial fishing in BC. I had been working with Indigenous people in the interior of BC to record their history, so I took a similar approach, making audio recordings. Equipped with a tape recorder and notebook, I was able to contact some of the legendary skippers of the twentieth century. It was great fun.

These were the skippers that I knew from the fishing grounds, and I selected them for their reputations as "highliners." Many top BC seine fishermen had come from Croatia, and I visited some, like Louie Percich and Dick Anzulovich. I had known them both from a distance as respected owners and skippers of big, modern double-deck seiners. Their stories were like many others. They had jumped ship from a deep-sea vessel visiting BC in the 1920s and got jobs on Croatian-owned fishing boats. When these men left the Vancouver Harbour and passed through First Narrows—in the days before the Lions Gate Bridge was built—they found themselves in new waters. But if the BC coast was unknown to these men, it was similar to the island-studded Adriatic, where their ancestors had fished for centuries, and offered potentially rich fishing with no ancestral connections or limits. It was natural to follow other immigrant fishermen to learn how and where to catch fish in these unfamiliar waters, from Vancouver north to the Alaskan border and out to Haida Gwaii.

Norwegian immigrant fishermen, accustomed to their home waters in the wild North Atlantic, tended to venture farther, going out west to the Gulf of Alaska and on to the Bering Sea to long line for halibut. While some Indigenous fishermen went as crew on the halibut boats, I did not meet any who skippered their own boats to those distant waters. Herb sent his boat out with a non-Indigenous captain to fish halibut in the spring, even though he had fished waters on the outside of the BC coast.

One of BC's most successful fishermen, Fred Kohse, spent the years between the wars helping on his father's homestead in a little bay just up the Vancouver Island shore from Herb's home fishing grounds in Johnstone Strait. Fred told me that one of his formative memories was of a seine boat that came into the bay in front of the family farm

Seiner *Kristin Joye* on a set in Johnstone Straits. Built by Gooldrup Hulls Ltd. in 1990 and owned by James Walkus in 2010.

to make a big set on dog salmon. Fred left the farm and rowed a skiff north to Rivers Inlet, where he learned how to gillnet salmon. He would go on to build several, always larger, boats and fish everything from Bering Sea halibut to South Pacific tuna. When I visited him at his house in Vancouver's exclusive Shaughnessy neighbourhood, he explained to me that he excelled at halibut fishing. This was because when he built his first good boat, he "went into Prince Rupert and hired a Norwegian skipper, and then I went as an 'in-breaker' on my own boat to learn how. I didn't do that for seining salmon, and I didn't excel in the same way."

Over the years, I have visited those source countries and met fishermen who stayed home on the Dalmatian, Norwegian, Japanese and other coasts. In their home countries, they went on fishing their ancestral waters, just as Herb preferred to do. Like salmon, homegrown fishermen tend to follow ancient habits as long as they remain in their natal waters, but relocate them to new territory, especially wide-open new territory, and they may become much more innovative and adventurous.

One November, I went into the fishing company's Vancouver offices to pick up my final settlement. The manager, noting how small the cheque was, asked me how it was going with Herb's young sons on the boat. Even then, I understood that the manager did not understand that Herb's responsibilities to his family and its traditions outweighed any demand for more production.

Louie Percich fished for the same company. They lent him money, first to buy a boat, then later, when he had paid back that initial loan, to build a new, larger boat. Within two or three years he would catch enough fish to pay off each loan. They would then offer him another loan for a still bigger boat. This was to keep competitors from luring away one of their highliners.

Another class of skippers that I came to know were those, like Fred Kohse, who came with open minds, from non-fishing families. Byron Wright won a reputation as probably the most aggressive seine skipper on the coast. The son of an Anglican minister in Alert Bay, he began as a young crewman with a local skipper. Over the next few years, he spent time with other skippers and absorbed their local knowledge. Using that as his skill base, he was soon skippering then buying bigger and better seine boats.

Unconstrained either by tradition or propriety, he built a series of powerful aluminum seiners, rigged them with heavy gear and proceeded to set in all the difficult and quasi-proprietary spots. In the late 1980s he had the big aluminum *Prosperity* built by Shore Boats. He made a lot of money in the herring-roe fishery, which was very competitive and often involved several boats crowded together and setting on a large school of herring. He explained to me that a very large steel anchor mounted on the bow like a ram, together with a very loud horn mounted on the mast, was enough to intimidate most of the competition. He did actually ram boats and once lost his anchor in another boat's rigging.

In the good old days, the companies provided services for their vessel-owning skippers as well as for their own fleets. Canadian Fish had Sterling Shipyards in Vancouver Harbour; BC Packers had Celtic Shipyards on the north arm of the Fraser River; and Nelson Brothers had their Queensborough Shipyard on Lulu Island, in the Annacis arm of the Fraser.

For each of the over 13 years that I seined salmon with Herb, we spent at least a week at the Queensborough Shipyard getting the boat ready. Each June, we would go to the boat at its winter moorings in the fresh water of the Fraser River. It was all exciting for a small-town boy from Campbell River. I remember driving, practically floating, through

Queensborough Shipyard in 1993, an important facility for maintaining BC's commercial fishing fleet. It was shuttered and the land redeveloped not long after this photo was taken.

the flat delta farmland in Herb's big blue 1959 Oldsmobile 98 with Herb Alpert and the Tijuana Brass on the radio. Like life in the movies.

The shipyard had its own marine ways, a big warehouse of boat supplies, woodworking gear, machine shops and lots of boats. There was even an ornate former North Vancouver ferry for storage. Dozens of company and private boats were moored to the floats and available for us to explore.

We would eventually get to the DB *No. 3*, which had lain idle all winter with little or no heat. There were often greasy engine parts on the galley table. The iron stove top would have rusted over the winter. In the fo'c'sle, the mattresses would be damp and musty. Mitzi would go to work in the galley. Herb would start the engine and let it idle some life and warmth into the boat. Soon Mitzi would have the stove top and galley table cleaned up. We would pack some groceries down the dock from the car. The boat would come slowly back to life as a living, breathing organism that would take us all, once again, up the coast to meet the salmon. Of course, we had to paint it first. I would go up to the shop and charge a new set of 9-foot oars to the boat's account. The boat would be hauled on the marine ways, and shipwrights would change the all-important zincs, blocks of solid metal attached below the waterline to stop electrolysis from eating away at the propeller and other metal parts. Any work was charged to Herb's boat account, which seemed to be limitless. One of the final chores was to repaint the name

The *Departure Bay No. 3* on the hard at Queensborough, getting spruced up for another season.

The Slide, a spot below Camp Point noted for its tricky tides but good productivity, is famous as Assu and Roberts family territory.

that carried the boat's history from her Japanese Canadian builders in the 1920s. Herb was very much of the belief that it is bad luck to change a boat's name.

After a week of this, Herb would turn the boat down river, going out to the big south arm between little Lion and Don Islands, then on to Steveston for fuel before continuing down along the jetty, built after completion of the Panama Canal in 1914 to bring larger steamships to the river for canned salmon. Finally, out past the Lightship, as the stationary light at the end of the jetty is known, before turning toward the mass of Texada Island and beginning the first leg of the 12-hour trip to Cape Mudge village.

The companies in the 1950s and '60s understood their fishermen and managed them to specific ends. In Herb's case, it was to fish that famously difficult but productive spot known locally as the Slide. It could be fished on both the turn of the flood and the turn of the ebb, but each tide had its own unique character. Like his father and grandfather, Herb knew how to read those constantly differing tides.

As a tie-up man, I knew and feared the power of these tides. On occasion they would turn faster than even Herb anticipated. When this happened, I would see the water being slowly wrung from the manila fibres of the beach line. I would look at the drift log that we had jammed into a crevice in the rocks, hoping that it would not come loose or break. At this spot or another, the weight of the long, deep net would be too much, the rope would break with a sharp snap, and the metal shackle that joined it to the net would come back like a 1-pound cannonball. I heard of multiple cases where an incautious beach man was killed when this projectile hit him on the head. I saw instances when it flew back into the salal that lined so much of the beach, cutting a track through the tough bush. At such moments, there was no time to recover the flying hardware. It was a rush to the skiff. The skiff man rowed us after the broken end of the line still attached to the net. We would retrieve this as the seine boat completed closing the circle of the set. In every instance that I recall, we managed to save the set before the seiner, like a dog catching its own tail, got into a real tangle.

The late Al Wagner started with Herb Assu in the 1960s and over his lifetime became a valued crew member and engineer on the family's boats.

The set at the Slide was legendary for its challenges. Years ago, a massive pile of gravel had come crashing down a mountain creek and splayed out into the strait. We sometimes set off an old piling that loggers had put there, but the best set was farther into a bay formed by the gravel and a rocky outcrop. Herb and Mitzi were each the oldest or next-to-oldest in their families. By the time I went fishing, Herb had taught several of his brothers-in-law to make the set, though they continued to leave that spot to him and generally set a little farther down the coast at Humpback Bay.

Herb didn't only train skippers. I realized early on that, as a son-in-law and not a son, I wasn't a potential skipper. I loved being part of a competent fishing crew in a way that I had never felt with team sports in school. The sense of belonging and of being an athletic part of a smooth-operating team felt good. At the same time, I realized that I didn't want to be an over-aged deckhand. Those I saw on other boats didn't look very happy. Accordingly, I began spending my winters in school, where I pursued my studies with a maturity and dedication learned on the boat—unlike the way I approached my high-school studies.

I had been fishing with Herb for only a few years when Vicki's sister Judy married Al. He was a wiry teenager as well, but he had learned to work on a carnival, with long hours and a do-everything ethic like that of fishing. He stayed with us for only one or two years before, having learned the basics, he took a job on one of the other family boats with a more aggressive approach to the fishery. Al built a strong reputation as a crewman and learned to be a fine marine engineer, a skill set that is always welcome on a boat. He worked at sea for the rest of his life without becoming a skipper but earning a well-respected place in the family.

I have often thought of a seine-boat skipper as analogous to the captain of a square-rigged sailing ship. Only, rather than looking up to see how the winds fill the sails, the seine-boat skipper had to look down to see how the tides shaped his net. Seine nets are built by their owner to include a "percentage," which refers to strips of excess web that cause the net to bulge outward creating a pouch that deters the fish from escaping downward before the bottom of the net can be closed during pursing. If not fished properly, this extra web could result in a big roll up of web along the lead line. These "bears," as we called them, could take a lot of time and energy to untangle on deck. Herb seldom got roll ups. The amount of "percentage" to build into the net and where to place it was a subject of conversation among knowledgeable skippers. Roll ups were not the only problem; a poorly designed net could also cause a backlash when spooling off the drum. Herb was a very good net man, for which I was thankful when I saw other crews struggling with tangled gear.

The tide changed at different times along this home stretch of shoreline. Often, after the set at the Slide, Herb would run the boat down along the shore. On the way, we would pass other Cape Mudge family boats at Humpback Bay, where we sometimes set, or at Little Bear River or at the Pineapple. We rarely set at the Pineapple, which had a good back eddy that often held a half-dozen or more big spring salmon. As we passed, I always admired the Dick family's big 1950s Matsumoto-built *Eva D II*, which, along

Herb's uncle Harry Assu was owner and operator of the *W No. 8*, seen here on a set at the little back-eddy bay known as the Pineapple.

with Harry Assu's immaculate *W No. 8* and a couple of other local boats, held a proprietary tradition in that spot. Later in the year, someone would repeat a local truism, "When the leaves on the big maple tree at the Pineapple change colour, it is time for the dog salmon to come."

A little farther down the Vancouver Island shore, Ripple Point marked the southern limit of our home fishing grounds. Across from Ripple Point was the entrance to Blind Channel, between East and West Thurlow Islands. In the fall, we fished dog salmon there. Ripple Point was at the base of a mountain that rose steeply from the shore. On summer afternoons, a strong westerly wind could come up, making it difficult to hold the skiff near the beach. Sometimes when we were hauling in and we had a big swath of web dangling from the power block, the wind would catch it like a sail and twist the boat around so that we risked getting wrapped in the net. Unable to use power to straighten the boat around for fear of fouling the propeller, Herb would calmly manoeuvre the boat back into position by getting the crew to manually pull some net aboard at the bow or stern.

If it was early in the day when we got the net back aboard and put whatever fish we caught in the hold, Herb would turn the boat back up toward the Slide. Often we would stop a little short of the Slide in a little bight that was known locally as Mitzi's Bay.

The *Rainbow Queen* was built in 1941 for a Japanese Canadian owner and confiscated a few months later. Here, the web hanging from the power block is catching wind and twisting the boat into the net, forcing the skiff man to pull in a lot of cork line to straighten the boat out. *Vicki Assu Robbins*

CHAPTER SIX

Grounding

It was at Mitzi's Bay that I experienced the only grounding of my maritime career. On occasion, when Herb knew that there would be a good tide at sunrise and the weather was favourable, we would anchor for the night in this spot. I'm not sure what happened that night, perhaps the anchor dragged a little or perhaps we let out a bit too much line. Regardless, about an hour before dawn, I heard Herb starting the main. As he came back through the fo'c'sle, he told the crew to get up. I could already feel that the boat had taken a slight list. Scrambling out of my bunk and into my pants, I hurried through the engine room and up through the galley to the deck.

Herb was at the anchor winch. I took up my usual place with the bar to guide the cable onto the drum. The boat's bow swung out in the direction of the anchor, but the stern remained firmly grounded near the shore. Herb was hoping that the anchor would pull the boat off the shore, but instead the anchor came dutifully up to the bow.

Herb then directed me to get my skiff and bring it to the bow. Once I was in position, he lowered the heavy anchor onto the back seat of the skiff. I was directed to row out into the strait. Once I was well out and the anchor line was at a lower angle to the bow, he called for me to push the anchor over. Herb was unwavering in his understanding, determined that the solution to a boat's problem should be found on the boat or in the surrounding waters.

The second attempt to pull the boat off with the anchor was no more successful than the first. As I got back to the boat, which, with the falling tide, had taken on more of a list, we were startled by the blinding spotlight of a passing tug. Once again, I was directed to row out into the strait to ask the powerful steel tug for help. I remember being pumped with adrenaline, but I was quickly deflated when the tugboat captain said, "I can't help you until I make arrangements with your skipper."

The "arrangements" were clearly a reference to salvage money. Herb came from a maritime community where one helped one's fellow mariner without expectation of recompense. He quickly waved the tug off. Our boat continued to grow a starboard list with the bow pointing out from shore.

Herb directed me to take a long line and row ashore, where I was to tie it as high up as possible on a big cedar. After I scrambled through the salal and, with the help of one of the crew, got the line secure about 6 feet above the ground and probably 10 or 12 feet above the water, Herb took the other end of the line and, wrapping it around the deck winch's capstan just behind the cabin, began to pull the boat so that it lay broadside to the beach.

Maintaining his calm and methodical manner, he kept tightening the line. The boat listed more heavily to starboard until it seemed that the top of the mast would touch the trees. The water was coming up on the deck and lapping against the bottom of the galley door sill when he finally tied off the end of the line. The swell of the hull was now cinched so tightly against the gravel ledge that there would be no risk of it toppling over as the tide went out.

The *W No. 4* was built in the US and brought to Canada around 1912. She sank in Seymour Narrows in the 1980s.

Daylight had broken by the time I got back on the boat. We all sat or stood on deck waiting for the tide to go down then come back up to refloat us. The big, old war-surplus cathode-tube radio-phone was on, and the usual morning chatter was broadcasting on the 2318 channel. A powerful transmitter, identified as the tug that had passed earlier, came on, "Attention all boats, there is a seine boat grounded about five miles below Camp Point."

Herb, who had been calm and steady throughout the ordeal to this point, suddenly sprinted across the deck and up to the wheelhouse, where he went online to assure the fleet that he had everything under control. While low-key and limited in his use of the radio-phone, he wasn't about to have some tug broadcast to the fleet that Herb Assu couldn't handle a crisis.

After the tide turned, the boat floated on an even keel again, and we could see from the skiff down to the propeller. Our underwater net guard, called a beaver tail for its flat, oval shape, was bent up into the propeller blades. Keeping things in the family, Herb called his brother Ronnie to come and tow us away from the shore. In deeper water, we again dropped the anchor and, from the skiff, studied the damage to the beaver tail. The pipe had bent only about a foot upward into the propeller. Herb sent me ashore to get a couple of boulders, which he then tied up in some old netting and attached a rope. The net guard was about 7 or 8 feet underwater, but, with a pike pole, we were able to wrap some line around the bent pipe. With the aid of the winch and the single-fall block on the boom, we could lift the bag of heavy rocks then let it fall into the water. Of course, the water reduced the weight of the rocks, but little by little we were able to bend the pipe down clear of the prop. The week's fishing was over. We made our way back south to Campbell River, where the boat was hauled out of the water and the net guard repaired. Fortunately, there was no damage to the hull, thanks to Herb's action in warping it against that harmless gravel beach.

CHAPTER SEVEN

End of the Cotton Seine

The first year that I fished with the Assu family, I was a newly married high-school dropout without a clue or a plan for the future. In 1960, the salmon season began at the end of June and extended up to November, when we finished up the season in the south, near Saltspring Island, fishing for dog salmon in Satellite Channel. I don't remember if we caught many fish, but I think that our cotton seine snagged the bottom and tore a long piece of the lead line from the web. Herb took it as a sign and said that we would call it quits since the net had to be stripped or disassembled anyway.

These were the final days of cotton seine; nylon web and lines had begun to come into use and were proving to be clearly superior. There were still a few Spanish corks around from the days when the net was hauled aboard by hand, but they were easily broken in the power blocks and had mostly been replaced by plastic floats—which are still called "corks" to this day. As we travelled out through Active Pass and across the Gulf of Georgia, I got to share in a disappearing fall ritual. We stripped the net of its lead line and cork line, then we cut the twine that laced each of the five, or perhaps seven, 100-mesh strips of web that had been laced together that spring.

After passing the Lightship at the end of the long jetty that marks the mouth of the main arm of the Fraser River, we ran about 12 more miles up the river to the old, inactive St. Mungo Cannery, at what is today the south end of the Alex Fraser Bridge. It was all new to me, so I had no idea where we were or that we were also near the site of an ancient and important former Sewqueqsen settlement.

In the late fall of 1960, the only part of the cannery still in use was the cavernous net loft. Reminiscent of a traditional coastal big house, this was one of many net lofts scattered along the BC coast, usually at deactivated cannery sites. In time I would work on nets in several, from Quathiaski Cove to Port Edward. Built with the finest quality BC lumber, they had vast smooth floors that the net could be dragged over when being built or mended after a bad snag on the fishing grounds.

The rafters were long, heavy timbers, strong enough to take the weight of the 100-mesh strips of cotton seine that we now lifted up and over. The cotton web had to be hung to dry or it would rot. Stowing the parts of the seine, cork line, border web, body web, lead line, as well as the bunt—a heavier mesh that held the fish to be brailed—was a hard day's work. It also proved to be Herb's last waltz with a cotton seine, as he followed the lead of other fishermen and purchased nylon components to build an all-nylon net the next year. Nylon seines could be left intact and piled for winter storage with no risk of rotting. They were also much lighter in the water.

Herb was happy to talk about catches of fish, but he was reluctant to talk about money. It was only from bits and pieces of talk by older crewmen, who also seldom

A troller and a seiner share the waters in Johnstone Strait.

talked about how much we were earning, that I learned of the share system for paying a salmon seine crew. Eleven shares were divided among the boat owner, the skipper, the net and the crew. The system had evolved from the days before power blocks, when there were typically eight men required to pull the net aboard by hand. The power block reduced that, so in round numbers I estimated that I made about ten per cent of the value of the catch after fuel and grub costs were deducted. When we delivered fish, a tally was kept on the packer boat, with a copy given to our skipper. Calculations to deduct fuel and grub were done in the fishing company offices. Since virtually everyone smoked in those days, we each got a carton of cigarettes on the grub bill from time to time. A "bonus" that we paid for in other ways.

The boat was given a "coupon book" for shopping in company grocery stores at places like Quathiaski Cove. I don't know if Herb ever did his own independent checking of the company's records. He also got all the repairs and winter work done at the company shipyard at Queensborough, on the Fraser River. My impression was that he was content if he could just keep fishing his bit of Johnstone Strait with his family and didn't worry about the finer points of company practices. Since he owned his own net, he got an extra share for that as well. I was told that in the early 1960s a new net cost about $6,500. Since it was paid the same as I was, I took that to be my cash value in the commercial fishing industry.

Prices and share distribution were negotiated by the United Fisherman and Allied Workers' Union (UFAWU) with the support of the Native Brotherhood. I was a member of the latter because I was on a Native boat. This created some difficulties because the union said that, as a white guy, I should be a member of their organization. I managed to avoid any direct conflict and proudly maintained my non-voting membership in the Brotherhood. I even carried the Brotherhood button fastened to the inside of my wallet for years. At the same time, I much admired the leadership of the union, especially its head, Homer Stevens. When we went on strike, it was made very clear to all members that the leadership would take no pay for the duration. The union's communist leadership always showed true respect for the fishermen and shore workers.

At home, my father got the UFAWU newspaper, *The Fisherman*, so I already knew a bit about unionism before I went fishing. My dues in the Native Brotherhood, along with those paid by other fishermen, funded some of their operation, but the union was active in a wide range of issues that varied over time. Always there was a struggle to unite the coastal and interior nations. The relatively affluent coastal nations were able to sell their fish, while the interior peoples were legally only allowed to fish for food. The Native Brotherhood published *The Native Voice* newspaper, which I also read. And then there was a small publication called *Fishery Facts*, put out by the company and referred to by fishermen as *Fishy Facts*.

CHAPTER EIGHT

Herring Fishing the South Coast

After storing the net at St. Mungo, we took the company-owned *San Jose* to the nearby Nelson Brothers shipyard in Queensborough. I don't remember how I got back home to Campbell River. I had not made much money over the four-month season, so I immediately began looking for work. A friend and I began cutting down old cedar telephone poles with a cross-cut saw beside an abandoned logging railroad grade. One of my few skills was splitting cedar fence posts for the fields at home, so I put that to good use. We had only worked for a few days when someone came up to tell me that Herb wanted me back on the boat for the herring fishing.

I was terrified and proud. I had been told that, since there were only about 70 licensed herring seiners, only those who were skippers or senior crew on salmon seiners would get a job. Nepotism won out, and I am eternally grateful to Herb for taking me on. In addition to me, the crew included Herb's brother, three of Mitzi's brothers, and two more relatives, for a total crew of eight. With only one year as a salmon crewman, I was the only one who was not a salmon seine skipper.

Herring fishing was very different from salmon fishing in that we usually fished nights, sometimes with huge lights to attract the fish, at other times with a paper echo sounder recording fish as they appeared below the boat. The herring were not travelling, so we didn't tow the net or tie on the beach. The skiff had a powerful gas engine and helped pull the net off the seine table. Herb set the net in a circle so as to arrive back at the skiff as the last of the net came off the seine table. I was assigned to the skiff with Mitzi's younger brother, Aubrey, who I knew from school. He was a few years older and had recently become a salmon skipper. Being closer to my age, he was always more helpful than critical of my learning.

The year before, the companies had paid around $13 per ton for the herring, which went into a reduction plant to become fish meal and fish oil. This year, the price dropped to $8.80 per ton, divided equally among the eight-man crew. If we got a full load, which we usually did, I could make, after fuel and grub, about $65. Not big money even then, but I was considered very fortunate to get a herring job.

I don't think that Herb had any grand plan for me, he just wanted his daughter's husband to have a job and earn a little money. A year earlier, I had been hopelessly ignoring my grade 11 teachers to play childish pranks or to read interesting novels under my desk. Now I found myself on a boat with seven master fishermen, all of whom seemed to be willing to teach me.

If salmon seining was elementary school, herring fishing was high school for this young white boy. The learning started quickly. The first night out, we anchored in the Gulf Islands and put on the bright lights. After a few hours of sleep, Herb had the crew stand by. Aubrey and I got in the skiff, and soon we were setting around the school. As we came back to the bow of the seiner, I threw a heaving line with a monkey fist on one end and on the other a line leading to the end of the net. It was a good-sized set and enough to load the *San Jose* to its 64-ton capacity. Fully loaded, the boat had an inch or so of water on deck, with only the raised combing around the hatch stopping us from swamping. With the hold full, the crew put the heavy hatch boards back in place. I had already been told that it is bad luck to stow a hatch board upside down when you take them off; now, with their recessed handles, they were placed back on. A brown canvas was laid over the hatch boards and then stretched in place by thinner batten boards that fitted into brackets around the hatch combing. Wooden wedges were then driven in to hold the tarp in place. Finally, 4-by-4-inch strongbacks were laid across the hatch boards and held in place with bolts. Our hatch was now securely sealed from any waves that would wash over the top.

Herb told Aubrey and me to take the big, heavy power skiff in and tie it to an old dock. I had no idea where I was, but I recall a weathered sign on the dock proclaiming it to be Otter Bay, which would put it on Pender Island. From there we travelled north up Trincomali Channel and out through Porlier Pass to cross about 15 miles of the gulf to the Fraser River and Steveston. It was a grey winter day with a stiff southeaster blowing up the gulf from Puget Sound. The steep, heavy chop and white caps in the gulf explained why we had left the heavy power skiff behind.

The crew had been up the whole night, and most of us had gone to our bunks well before transiting Porlier Pass. It was a chance to grab a little nap before we got to Steveston. Our sleep was interrupted by Herb's brother Ronnie. As the engineer, he had grown concerned when the waves began to wash higher over the deck. Looking aft from the galley door, he thought the deck seemed to have even more water than before. He went below and started the pump, but when he went back up and looked over the side of the boat, no water showed at the through-hull outlet. He went back to the engine room and saw water coming through the wooden bulkhead that separated the fish hold from the engine room. On a modern boat, this would be a watertight bulkhead, but the *San Jose* had been built in 1928, the era of the Model A Ford.

Previous pages: Steveston's "Cannery Row," taken from the deck of the *San Jose* in December of 1960.

Ronnie called into the fo'c'sle, and we all got up and went on deck. The man on the wheel had also noticed the bow rising, but he had simply moved the stool over so that he could see around it. There was a big cast-iron hand pump on deck just aft between the hatch combing and the net table. You pumped with a long steel handle, and the rubber diaphragm brought up nearly a bucket of water with each stroke. We all took turns pumping. Usually when you get on the pump handle, everyone lets you stay there, but on this occasion, after a minute there was another crewman offering to take over. I pumped for a bit, but soon the water on deck was about to run into my low boots.

We had a dead skiff, like the one that I had rowed on salmon, but there were no oars, and as soon as we launched it, it was swamped by the southeaster that was blowing up our stern. I looked at my mentors and noted that there was no bravado or joking about our situation. Someone suggested that I go down and get the old kapok life jackets from the fo'c'sle, but recalling the water coming into the engine room, I declined. No one went. We just kept pumping. Soon a green-blue colour from the copper sulphate that we used to preserve the net could be seen coming up through the bilge pump. "We are taking water in the lazarette," someone said, as the sacks of blue stone for the cotton seine were stowed in that separate stern compartment.

Herb had taken over the wheel. The crew, all standing on deck, speculated that if we made it to the Fraser River, we might sink because the boat would be less buoyant in the fresh water. Still no joking among my seniors. I was truly scared as we came into the river at the Lightship and began making our way up beside the long stone jetty to Steveston. In Steveston, as we ran up along Cannery Row to our plant at Paramount, it felt safer but still seemed that we might not make it.

Later, I heard from a friend on another boat that the fleet had been following our progress. He had heard Herb come on the radio calling the plant, "We're taking some water. It would be good to offload as soon as possible," he said in his usual calm voice. No alarm, just saying that we were sinking.

The alarm and anger came when the herring and water had been pumped out and the crew went to inspect the planking around the stern where it has been leaking into the lazarette. The dried caulking between the planks had been pushed into the hull by the water pressure, leaving gaping gaps between the planks. The hold had not been filled all summer, and the sun and heat had dried the planks and the caulking. The crew's anger was directed at the company for not properly inspecting and repairing the boat before sending us out on it. The pump had failed because an elbow in the piping was found to have rusted through so that it was sucking air from the top of the fish hold.

The crewmen were all productive salmon seine captains who made a lot of money for the company. Their angry conclusion was that this lack of regard for their safety on the company's part was one more example of the racist discrimination that had long been a feature of the BC fishery and Canadian society.

Not only was the boat crewed by Indigenous skippers, but its capacity was also only 64 tons, half that of some of the larger boats that tended to keep year-round crews. At 35 years old, the *San Jose* was twice the age of many of the big double-deck seiners that we fished alongside. Later in the season, I would see Herb, the master herring fisherman, make sets of up to 500 tons of fish. Because of our smaller capacity, we were pooled with only the aging 1926-built *Ribac*, which also had an Indigenous crew and was skippered by Herb's uncle Thomas Assu. That boat was so old and waterlogged that they brailed fish in until it was near sinking. A full hold would have sunk it. After loading our two boats, Herb, always good natured, would invite some of the big boats to come up to our cork line and brail out the rest of the fish. This meant long hours of work handling the net with no return to our boat.

Above: In December 1960, Herb took Alan seining reduction herring on the *San Jose*, shown here with 57 tons—a nearly full load.

Left: Teamed with Herb's uncle Thomas Assu on the *Ribac*, the two boats pooled catches, which sometimes meant a break for the *San Jose* crew while the *Ribac*, as here, did the work of swinging the brail across its deck to fill the hold of the *San Jose*.

CHAPTER NINE

Up North

We fished the Gulf Islands into December and then went north to fish out of the Port Edward plant near Prince Rupert for a week or so before Christmas. It was there that I saw several good-looking salmon seiners with the company's green hulls. It was explained to me that these were built by Haida people, but then, due to debt incurred in their building, they were taken away from their owners by the company. At the time, my fellow crewmen saw this as further evidence of the mistreatment of Indigenous fishermen. Sixty years later, this is still a cause of pain and anger for Haida people.

I learned something about navigating a boat in the dark and snow without radar, but I also learned of the racism that the BC fishery was built on and the deep-seated anger that my fellow crew carried as a result.

The older men in the crew drove me to endure long days and nights, and they could be harsh. My father had given me an old grey wool mackinaw that I was proud of, but the deck boss saw it and asked sarcastically, "Where did you find that? It looks like it has been worn by a man."

Brailing meant using a giant dip net or brailer with an 18-foot handle and powered lines to scoop herring out of the seine and drop them into the hold. When accomplished with the right timing and teamwork, this was a beautiful operation. One man was on the skimmer line to pull the brailer through the net full of fish, another helped swing the full brailer when it was lifted, while the deck boss on the long wooden brailer handle directed the operation. It was exhilarating teamwork, and once the rhythm was set the boat rolled to starboard as it began to lift the 1-ton brailer and rolled to port as it swung across the deck. It felt like the boat was waltzing the fish aboard. My job was to trip the "asshole" line to let the fish out of the brailer as it passed over the hatch. I was told in no uncertain terms that if I missed, it would be me who would shovel the fish into the hatch, but only after being called "Asshole!" by the crew. I made certain not to miss, but even the threat added to the excitement.

Ribac coming on *San Jose* cork line during winter herring season. The *San Jose* could pack a maximum of only 64 tons. With sets of up to 500 tons, other boats tied on the *San Jose*'s cork line and helped themselves to the leftovers.

We fished the north coast for reduction herring in January and February, meaning these precious fish would be ground up for feed or fertilizer when their oil content was high. Years later, the oil would be consumed as the maturing herring developed roe or milt, but the roe fishery, with its $2,500-per-ton prices, had not yet developed. The roe fishery opens in March, when the roe in the females has grown to over 10 per cent of their body weight.

It was on the herring that I was introduced to the power and beauty of the coast's fleet of "double-deckers." These jumbo-sized boats had a second pilot house up top with a walk-around outer bridge deck behind a windbreak or dodger. This was often called a "Portuguese bridge," in recognition of its origin among the Portuguese-American fishermen of southern California, who used them on sardine and tuna boats. Many of the early double-deckers were built in the US as sardine seiners, but when those little fish disappeared, they became surplus. They were too large for salmon seining in Alaska, where vessel size was restricted by law, so, many were bought by Canadian fishermen to intercept the Fraser sockeye in the Strait of Juan de Fuca before they passed through some US waters on their way British Columbia's Fraser River.

The boats were at first considered too big and cumbersome for salmon fishing in inside waters, but with their large packing capacity, they were excellent for herring fishing. In the 1950s, a lot of Prince Rupert–based halibut fishermen had such boats built

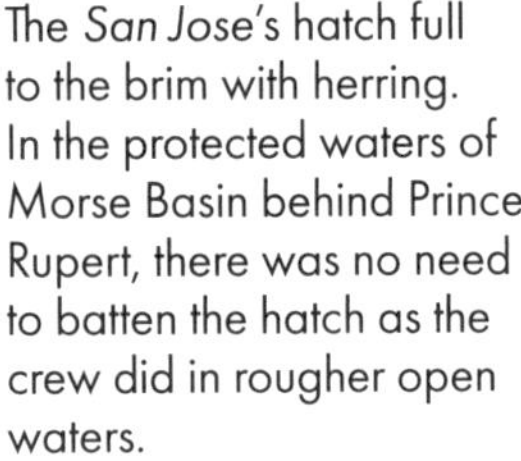

The *San Jose*'s hatch full to the brim with herring. In the protected waters of Morse Basin behind Prince Rupert, there was no need to batten the hatch as the crew did in rougher open waters.

Left: Built by Matsumoto Shipyards Ltd. in 1957, the *Silver Bounty* was on sea trials in this photo commissioned by Finning Tractor, who supplied the Caterpillar engine.
Commercial Illustrators

Below: Two boats working a single herring set. Alan took this photograph from the bridge of the *San Jose* in February of 1960.

VANCOUVER BC

Having made a big set, the *Western Mariner* (centre) brails across her deck to the *Sleep Robber*. The *Western Warrior* has come on the *Mariner's* cork line and is brailing while the *Amlac* stands by in February 1960.

for long-lining Bering Sea halibut, as well as for salmon and herring seining. They went to Sam Matusmoto, a Nikkei boatbuilder who had been interned during the war and finally released in 1949. Sam and his dad had built boats—and a fine reputation—in Prince Rupert before the war. After their release, they established a yard on Vancouver's North Shore, where they built fleets of 34-foot gillnet boats and some single-deck seiners, like the Cape Mudge-based *Eva D II*, but most of the northern fishermen soon wanted big double-deckers. I can still vividly recall the night that the gleaming white hull of one of these vessels emerged from the darkness and a light fog to show her massive hull and huge bows bearing her name, *Silver Bounty*, in strong black letters. She was like a dream. I thought that if there were boats in heaven, they would look like this.

As much as my fellow crewmen taught me, the waters and the winter fishing in Hecate Strait made an even stronger impression. When I first went out, I got seasick just standing by in the skiff as the swells pulled it away then banged it into the stern. My skiff

Tony Roberts shovelling snow on the *San Jose*. He was a gruff deck boss with little patience for a green white boy, but he taught clear lessons.

mate, Aubrey, showed kindness by ignoring me as I puked overboard. I quickly learned about Gravol. Each night I took one of these amazing little pink pills as we approached Eddie Pass on the way out to the "edge" where the herring showed, and I was fine for the night.

I came to like the fishing, especially the point where we came back to the seiner in the skiff. The big boat would be riding up on a swell, with its green bow and the copper bottom paint showing, before chopping like a huge axe back down into the next swell. I would have to time my throw of the heaving line to land it on the forward deck. It felt like real seamanship to a boy who had grown up in sheep pastures reading novels like Charles Kingsley's nineteenth-century *Westward Ho!* or my own father's more contemporary *Saltwater Summer*.

The Gravol and the long hours did make me sleepy. One night, the chain that drove the big mechanical deck winch from a shaft that followed the deck head aft from the main engine broke. Our forward hold was already full of fish. However, by opening the small aft hold, there was a space created on top of the fish. I was selected to lie on my back on the load of fish and slide up under the deck until I could pass the broken chain up to the guys on deck.

There is always a good swell running in Hecate Strait, and this night was no exception. The mass of fish rode up and down the sides of the hold as the boat rolled and individual fish flipped and gave my legs and body a gentle massage. The crew on deck told me to wait for a bit before passing up the end of the broken drive chain. I was sleepy. It made for a good story, at my expense, in the Rupert bars that weekend. There are still people from among that crew that recall me falling asleep on top of those fish. However, I was not, as some have claimed, clutching a single erect herring between my legs.

Surviving and learning to pull my weight on that boat did a lot for my insecure teenage confidence. My daily teachers were my older crew mates and family. Herb tended to spend more time in the wheelhouse and leave the deck to his knowledgeable crew. There were three related events that I recall in February of 1961. I'm a little unclear on the sequence. A huge storm caught us in an insecure anchorage among the islands on the mainland side of Hecate Strait. As usual, I didn't know exactly where we were, and Herb decided to move to a more protected area. This required that we run north for a distance in the storm-lashed strait.

The wind on the side of the boat's cabin kept it rolling from an upright position to a harsh angle to starboard. I was sitting in the galley with some of the crew, who were holding a pencil to show the angle of our roll and predicting that we would be OK until it exceeded 45 degrees. I wasn't too worried because the people who knew about such things were still laughing.

Then the deck boss came into the galley and said, "You better go and check your skiff." I got up and looked out the top half of the galley's Dutch door. Every swell was cresting over the port side and crossing the deck in white foam. I was more afraid of the deck boss than the ocean. Waiting for a space between the swells, I dashed out the door, went hand over hand along the skiff's heavy painter tied to the deck winch and climbed up on the massive cotton herring seine. With my fingers laced in the web, I crawled back so that I could see into the power skiff being towed, snug to our boat's stern.

The skiff was half full of water. I turned to look back across the deck and saw Herb leaning out of the top half of the Dutch door furiously shaking his fist at me. I was shocked and wondered what I had done now. Quickly making my way back across the deck, I went into the galley. "What the hell are you doing going out on deck in this weather?" Herb asked. "Don't you know these guys were just teasing you. Even if the skiff is full of water, there is nothing you can do about it!"

Somewhat chagrined but also proud of my stupid daring, I proceeded to enjoy the tumultuous travel of the boat. I think that it was somewhat later in the same storm that Herb let me take the big wooden wheel. As I recall, the huge swells were quartering on our stern, which can make for a difficult job of handling the boat. The boat was twisting and corkscrewing its way up and over huge waves, and the wheel had to be shifted to compensate. I did well for some distance until, on a particularly large roll, an empty coffee cup bounced off the compass table. Without a word, Herb took the wheel back. I had been given one more wordless lesson, I like to think, as a reward for my mindless adventure on the aft deck.

The ill-fated *Northview* in Vancouver Harbour at a time when the Vancouver Hotel, CPR dock and Marine Building were dominant landmarks. *Commercial Illustrators*

We got into good shelter, but later we learned that one of the herring fleet, a big double-decker called the *Northview*, had gone down in Fitzhugh Sound. The news also said that the whole coast was being blasted and that the storm winds had blown out the windows in the Eaton's department store in Vancouver. I can't find any record of this storm or the sinking of the *Northview* online. My memory wants to put the winds at 80 miles per hour.

My own experience of the storm was punctuated by our participation in the search, in post-storm calm, for the *Northview*. Hours spent staring into those cold green waters gave a teenage mind time for reflection on the vulnerability and potential horror that we lived with. I believe all the bodies, mostly from the owner's Croatian family, were found.

Many winters later, I travelled north through Hecate Strait on a large cargo ship to research a story that I was writing on the BC coast pilots. I looked out from the comfort of my cabin high in the stern of the ship to observe the sea, largely white with spindrift from the wave crests—Beaufort language for "Strong Gale to Storm." The weather report said that it was blowing over 50 miles per hour. It was at least that on the night of the storm in February 1961. But now, a half century or more later, I was on a ship that merely exhibited a comforting rolling motion as it moved up the notorious Hecate Strait.

It was shortly after that 1961 storm that we got our last load of the season. While Herb had allowed other boats to load up from his net several times, this was the first time we were able to load up from another boat's excess catch, down in Laredo Sound. After a final delivery to the Port Edward reduction plant, we headed south. We took a few days off in Campbell River before taking the boat back to the shipyard in the Fraser River.

Later that spring, I nearly got to go eulachon fishing in Knight Inlet. The net was ready and piled on the float at Cape Mudge. I'm not sure why Herb decided not to go, but I do understand that there are extensive protocols around participation in the fishery. Since it is not a commercial fishery for non-Indigenous fishermen, the government thankfully leaves the protocols of family and stewardship up to the Indigenous community.

From the outer west coast of Vancouver Island, the Nuu-Chah-Nulth fishermen, with their whaling tradition, readily joined sailing ships going north to hunt fur seals and sea otters. I heard stories of men who had travelled on those ships all the way to China before returning to their homes. Herb's ancestors had entertained no such far-reaching expeditions. Before the colonial commercialization of their fisheries, they had harvested most of the salmon in the estuaries of their natal rivers. Autie once took me into her smokehouse near the beach at Cape Mudge. She complained that the chum salmon she was smoking were too fat and would have to be canned. "The boys bring these off the seine boat," she said. "In the old days we would catch them in the river. Then they are not so fat and they smoke and dry better."

For a number of years around 1960, Herb sent his boat out with a white skipper. I think that the boat made more than one trip to the grounds off the northern BC coast and then delivered to the Campbell Avenue fish dock in Vancouver. One year, the halibut skipper for some reason asked Herb to take the boat down from Quathiaski Cove to deliver the final load. We took off the halibut gear, including the big horizontal wheel known as a Christmas tree and the chute for setting the long line, and stored it in Quathiaski Cove. Then Herb and I, with his two teenage sons, Harvey and Wayne, took the boat down to Vancouver, where we delivered the load of halibut at Campbell Avenue fish dock. Campbell Avenue was a marvellous U-shaped set of piers with individual independent companies renting small processing facilities. It provided excellent fresh fish to Vancouver's restaurants and markets until the 1980s.

It was also the location of the famous halibut board. Not a boardroom full of stuffed shirts, but an actual board, a simple chalkboard with the vessel name and listing of the sizes and quantities of halibut delivered. Buyers could then simply chalk up their offering price. This was one of the few truly public pieces of the BC fishery and was well known on the coast. "What's the offer on the board?" I would hear fishermen ask one another over the radio-phone. The first loads of the season got a premium price, but no matter what the price, it was democratic and public, like no other aspect of the BC fishery then or now.

Campbell Avenue was also the home of the famous Marine View Café, which was, until the whole dock was demolished, considered by many to be the best restaurant for fresh seafood in Vancouver. Its loss is still lamented by those who were in the know. It is generally understood that the Campbell Avenue fish dock was expelled by the voracious appetite of the Port of Vancouver for expansion of their container piers and grain silos.

Unloading a hold full of slimy halibut in the 1960s was my introduction to that magical and much-lamented place. With the fish delivered, we ran back out under the Lions Gate Bridge and up the main arm of the Fraser River to the Queensborough Shipyard near New Westminster.

Herb and I were tired from travelling all night, delivering the halibut and coming up to the shipyard. The boys had slept for a good part of the trip. Herb tended to let

The Marine View Café at Campbell Avenue's fish docks was famous for bulging shrimp sandwiches, fresh-caught fried sole and other prime seafood from the docks. And the view was great.

things slide for the younger kids, but now they wanted to go into New Westminster to see a movie. Ever indulgent, Herb dropped them at the movie, then he and I retired to one of the many bars on Columbia Street. There I would get another of Herb's lessons in proper deportment. I could see that he was dead tired and had hardly touched his glass of beer. "Why don't you just have a little nap," I proposed. "I'll wake you up when the kids get out of the movie." My mild-mannered skipper replied with a mix of anger and some deeper annoyance that I was not, at the time, familiar with. "I have never fallen asleep in a bar!"

I had seen house parties in Prince Rupert raided by the RCMP, who accused some of our lighter-skinned Indigenous crew of selling alcohol to "Natives," while the Indigenous house owners were accused of illegal purchase of alcohol (the Indian Act prohibited the sale of alcohol to First Nations up until 1985). I came to understood how the complex regulations regarding alcohol had become one more example of legal racism. The result was that Herb feared a prejudicial reaction from the bar staff if he dozed off, even if from sheer exhaustion. Without a long explanation, he had given me another lesson in Canada's institutional racism.

Herb maintained strict rules for public decorum. Once, about 1963, I was in the line-up for one of the only public pay phones at Ocean Falls' Martin Inn. Everyone wanted to phone home and, as it happened, I was in the line just behind Herb and Mitzi. Behind me were a couple of white guys who had just come up from the bar. This was before the late 1960s' change in societal rules around swearing in public. One of the two inebriated fishermen behind me declared loudly that something was "a big fuck-up."

Mild-mannered Herb roared, "You have no business using that kind of language in front of a lady! Now apologize."

I was not feeling nearly so gallant. My thought was, "Damn it, Herb, am I going to have to join in if you fight them?" Fortunately for my timid self, the two boisterous fishermen apologized for their language. Herb was always a skipper.

As a teenager in those times, I was often a bit challenged as how best to fit in. I recall another house party in Prince Rupert where, after the bar had closed, we gathered for more drinks and more fun. Someone brought out a broomstick, or perhaps a hockey stick, to serve as a ceremonial talking stick. This was passed from hand to hand and speaker to speaker among the men. In loud ceremonial voices, each speaker, if he was local, proclaimed his tribal and familial identity and welcomed the guests from the south.

One of our crew would then take the talking stick and, thumping it on the floor, proclaim their tribal affiliations and formally thank our hosts for their great hospitality. At some point, the talking stick was passed to me. I stood as tall as I could manage and introduced my tribal affiliations by saying, "I am the only white man in the house!"

I was chastised the next day by one of our crew, who teased, "Don't ever say that again. You are lucky that we got you out of there alive!"

CHAPTER TEN

Bootlegger Al

A year or two later, I had another encounter with alcohol in the fishing community. Herb often hired an uncle, Jack, on the family boat. In 1963, there was word of a particularly strong return of pink salmon in the central coast around Namu. We fished one week down around Addenbroke Island lighthouse. I think that we had a three- or-four-day weekend. We ran into Namu for a night, then went up to Ocean Falls for groceries and some R & R.

Jack had brought his friend Fred to round out the crew. Fred had spent the winter logging and still had a nasty leg injury that was becoming infected. At Ocean Falls, he checked with the hospital for treatment, and they admitted him. Jack and I went to visit, and Fred explained that they were going to keep him in the hospital for a week. Fred asked me to pick up a couple of bottles of whisky for him. He told me to hide one on the boat and bring the other to the hospital. We all thought this was quite fun and even a little funny, as it was well known that my father was the magistrate in Campbell River and that Indigenous people had only, the year before, been allowed equal access to the liquor store—a change that my father had strongly lobbied for, as the ban had been clearly discriminatory.

I did as Fred requested, with a joke about being a bootlegger. All would have been OK, but Jack and Fred got drunk in the hospital and were thrown out on the Sunday morning. They were reasonably sober when they came down to the boat, and I was pleased that we wouldn't be short crew.

It is a three-hour run from Ocean Falls down to Namu, and we went on for some distance south of that. Jack and Fred spent the whole voyage in the fo'c'sle. It was a pleasant Sunday cruise, and as the 6:00 p.m. opening approached, Herb told me to go and get those other two. That was when I realized that they had spent most of the afternoon finishing off the second bottle, the one that I had hidden on the boat for Fred.

Fred was fairly high, but Jack was plastered. He came on deck but fell trying to put on his slicker pants. Jack was the beach man, and Herb, still unaware that they were drunk, yelled at him to get in the skiff. We still had a table seine, so, as he was a small

The fleet at Ocean Falls in the summer of 1960. The *San Jose* is next to the float in the second row with the *W No. 4* next to it.
Royal BC Museum

man, I helped him climb up over the net and kind of threw him into the skiff.

There were so many pink salmon returning that the company had announced each boat would be limited to only 3,000 fish that they could deliver to the packer. Herb had located us in a good spot for our first set. At 6:00 p.m., the Fisheries patrol boat announced that the fishery was open. Herb turned the throttle up, and smoke came from our stack as the *DB No. 3* approached the beach. Herb waved, and Fred let the skiff go.

The net started off the table, and I rowed the skiff ashore. Fortunately, there was little or no tide. Jack climbed over the bow and draped the beach line over a couple of rocks. There were several jumpers inside the cork line, and Herb closed up quickly. When he waved us to come back to the boat, I managed to drag Jack, who was still very drunk, back into the skiff.

I rowed back to the seiner and jumped up on the bow of the boat to go around the cabin and pull up the gable end of the net. Jack seemed to be tying the skiff to the bow, as was our practice, and then he came back and began pursing one end of the net while Fred pursed the other end. Herb was plunging when he saw our skiff drifting away in the wind. Furious, he yelled that the skiff was adrift. He had figured out the inebriated state of half his crew and was getting agitated, as we never drank on the boat. Jack often crawled his way onto the boat on a Sunday morning, but he was a competent deckhand for the rest of the week.

There was a good breeze blowing and the skiff was already 60 feet down along the cork line. Jack, still coiling the purse line as it came off the winch, looked at me and said, "Could you take over here. In all my years, I have never lost a skiff before."

Next thing any of us knew, Jack had walked to the bow, kicked off his boots and dove into the choppy seas. Herb dropped the plunger and, in his usual calm under stress, directed us to take down a small skiff that was on the cabin top full of old lines and stuff. Once we had it in the water, he told one of his young sons to row it out and get the big skiff and Jack. It all took a while, and by the time we got the two skiffs back to the boat the set was lost because we had not finished pursing up, and most of the fish had escaped.

Opposite: Alan guides the brailer on the *DB No.3* while Wayne Assu holds the skiff away from the boat. *Vicki Assu Robbins*

No one got fired. We made a drift set with me alone in the skiff. Brailing was a struggle, as the two guys were still not sober. Somehow, we got our net and fish aboard, and we estimated that we had about 4,000. This helped when we got to the packer, as a boat that was short of the 3,000-fish limit came and helped us pitch our catch onto the packer in return for our extras. Somewhere in the confusion of that drunken set, Jack and Fred decided to put the blame on me and began to chant, "It's all Al's fault. Dirty old Bootlegger Al." The nickname was given special value and humour as everyone knew that my father was the magistrate in Campbell River. It spread in the family fleet and stayed with me for years.

A crewman rests on the net pile while the *DB No. 3* heads for the anchorage in the late afternoon sun. Long days.

CHAPTER ELEVEN

Seymour Narrows

I think 1963 was the last year Herb used a power block rather than a drum. A power block, which hung from a boom high above the deck so the net could be threaded through it and slowly pulled aboard, had been a great improvement over the previous method, which was to pull the net aboard by hand. The drum seine, which wound the net on a large horizontal drum at deck level, was a further quantum leap forward and allowed sets to be done much faster. Most other boats had converted over the past decade and the difference between what I would call Herb's gentlemanly way of setting and the more industrial drum-seine approach was evident when we delivered. With the power block, Herb had made four perfect sets with his interpretation of tides and currents, as well as generations of knowledge of salmon behaviours. Less experienced or knowledgeable drum-seine skippers used an industrial mentality to keep setting their net 10 and, in places with less tide, I've heard, even 20 sets a day. We would deliver 200 or 300 sockeye, and they might deliver 700 or 800. Individual sets might be less successful, but they had a larger catch for the day. It all reminded me of my father's teachings on fly-fishing. In his mind, the reading of the river, the currents, and even the surrounding natural bounty, along with a perfect cast, were more important than how many fish one caught.

Herb's son Darryl recently told me that the increased speed with which the net was hauled back aboard on a drum, compared with a power block, gave his father time to make four more sets per day, for a total of eight, without being pushed onto the rocks. But Herb preferred to continue making his four perfect sets in harmony with the tides.

The *DB No. 3* had been powered by a huge, deep-throated yellow four-cylinder Caterpillar 77. It was started with a little gasoline pony engine and then ran day and night with never a falter. There was enough surplus power to run the hydraulics required for the power block, but not enough to power a big, heavy drum that could wind in the whole net.

I expect that Herb had to take on some company debt to make the change over that winter of 1963–64. On the 12-hour trip up the coast from Steveston, with the new engine, all seemed fine to me, but not so to Herb. By the time we approached Seymour Narrows, just north of Campbell River, the tide had turned to flood. This meant that the water in the centre of the narrows would surpass the boat's speed and our ability to stem it. But the tide also sets up a narrow back eddy off the point of little Maude Island, beside Quadra Island. Herb ran the boat up along that back eddy, as was normal against a flood tide. After feeling a bit of the rip between the eddy and the flood, he throttled back and let the boat ease into the shelter below Maude Island, where we dropped the anchor. He directed me to go ashore and load up as many heavy rocks as possible. In due course, one of the boys and I brought these back to the boat, where Herb told us to pass them up over the bow one at a time. They were then lowered through the skylight and a trap door in the floor of the fo'c'sle. I would guess that it was several hundred pounds in all.

I don't recall if we brought out one or two such loads, but after two or three hours of this labour, Herb seemed satisfied. With the anchor up, Herb turned the boat out into the back eddy still flowing on the tail end of the flood tide. He angled the boat from the eddy into the still-strong flood. As usual, the boat rolled from side to side as he eased it

An American seiner, with its power block and power skiff, heading home from Alaska to Washington State through Johnstone Strait. Built in 1973 by Marine Construction & Design Co., the *Adirondack* was owned by Alan Jacklet in 2010.

across the rip and increased the engine speed. I realized then, as I had before, that such tidal navigation was not a learned science for Herb. It was a deep-seated, perhaps even ancestral, knowledge of how the sea felt as the energy passed up through the boat and through the soles of his feet.

Seymour Narrows is a defining feature of the BC coast. There are other tidal rapids, but this is the only one on the main transportation routes along the west coast of Canada that is used by virtually all the commercial traffic from the lower 48 states to Alaska. For captains, it is either a trigger of dread or adrenaline or both. No one takes it casually and survives. Before the top of Ripple Rock was blown off in 1958, it was even more of a hazard, especially for larger ships. It is thought US pressure played a role in getting the Canadian government to remove that hazard.

The passage for local fishing boats underwent little change with the removal of the rock. Those with the know-how learned to use the eddies to make it up over a hump of water that builds on both the flood and the ebb. I was often on the wheel as we approached the narrows. If it was a slack or mild tide, Herb would leave me to take it through. Southbound with a fair tide is relatively straightforward. At the bottom end of the narrows, the full force of tide crashes into Vancouver Island at Race Point. You must then simply follow the rip across toward Quadra Island.

Aubrey Roberts, Herb's younger brother-in-law, the one with whom I shared the power skiff on herring in the winter of 1960–61, is now retired after a long and successful fishing career. In the 1990s, I went northbound with him through Seymour Narrows against a strong flood. He took his big steel boat, the *Western Investor*, up the same narrow back eddy below the Maude Island light that Herb had used. But he did it sitting in a pilot chair, dialing the rudder on a knob atop the handheld control for his autopilot! A couple of years later, I asked him how best to make it southbound through the narrows when bucking an ebb tide.

That year, in August of 1998, Aubrey was fishing the massive 85-foot steel seiner *Royal City*. He emphasized to me that it was important to maintain speed when travelling through the tides: "You can't go less than about three-quarter speed. Once you cut back you have no control, because it is boiling. People go there and cut the throttle back to half speed, then when you try to turn the rudder there is no reaction, because you haven't got the thrust and you're pushing against those boils and whirlpools. I very seldom slow down from close to full speed when I'm bucking."

Coming south toward Campbell River against the ebb tide, the local boats stay on the Quadra Island side to avoid the stronger current on the Vancouver Island side. The tip of Maude Island, with its light, juts out a short distance, creating a back eddy on both the flood and the ebb. Southbound, the boats come up along this back eddy. "You can go within 12 feet of the shore, but we don't go that close, usually about half a boat length—around 40 feet off at the light," Aubrey explained, making it sound easy to move an 80-foot boat at 10 knots, plus the speed generated from the back eddy, between a rocky shore on the port side and a foaming tidal bore on the starboard side. Add to this a half-dozen 14-foot sport-fishing boats trying to catch salmon schooled up in the same eddy, and it gets even more interesting.

On the Quadra Island side (far side) are, from left to right, Deepwater Bay, Plumper Bay, Seymour Narrows and Maude Island, the latter joined by a causeway over Canoe Pass. In the foreground, on Vancouver Island, are Brown's Bay and Menzies Bay with log booms.

The excitement and rolls come when, just below the Maude Island light, Aubrey and other knowledgeable skippers angle their bows across the tide rip. This is a point where smaller boats that roll badly will cut the throttle to three-quarter speed just as they cross the rip and then crank it right back up against the strong green water. "There is a hump, a little kind of drop there, and you try to have enough pressure or speed to get over that hump," Aubrey said. "Once you're over the hump, you're clear. We've made it with a 9.5- or 10-knot boat against a 13-knot ebb, but you've got to hit the hump just right and crank her so you're skipping over there. You've got to know how to handle a boat, of course. A lot of times you think you're going to hit the beach, but there is enough pressure boiling off the rocks to push you off there. It's nerve-wracking as hell for people who are not familiar with it."

As Aubrey explained it, once you are over that hump in the water, it is just a matter of time. After getting up alongside the Maude Island light, he starts angling over toward the mouth of Menzies Bay, on the Vancouver Island side. "If you are halfway across and

The line tracing through Seymour Narrows shows a series of seiners following the route that a southbound boat takes to buck the ebb tide. This can put them in mid-channel when a northbound ship is transiting with a fair tide.

still holding your own," he remarked, "you're home free. Just watch the light, and as long as you're not going back, you're OK."

This is the course that puts the southbound fish boats in conflict with the northbound vessels, as the seiners pass slowly at a diagonal across the northbound shipping lane. While this route into the slower water in the mouth of Menzies Bay and then out around Race Point is not the most direct, it is the quickest way, as it avoids the strongest current. They can go close to Race Point, on the south side of Menzies Bay, because it drops right off to deep water there.

These local-knowledge directions for transiting Seymour Narrows come with the disclaimer that they are not recommended for skippers who have not had the kind of mentors and experience that Aubrey Roberts has had. Even those who have similar training are not immune to mishaps. Aubrey said that years ago his brother Gerry was coming up the back eddy behind the Maude Island light on the old *Sea Biscuit*. When he cranked it over to cross the tide rip, the steering chain broke and the boat swung onto the beach, knocking its bow stem out.

There is beauty in the way boat handlers like Aubrey Roberts and my skipper, Herb Assu, "feel" their way through the Seymour Narrows. Theirs is a proud and long-established tradition. Wise officers and pilots on northbound commercial vessels learn to anticipate the movement of seine boats as one more challenge of the passage.

Every BC fisherman knows of at least one seiner, like the beautiful wooden *Miss Joye*, that has either rolled over or sheared into the beach in the narrows. I recently asked a fishermen's Facebook group for names of vessels that had rolled in the narrows. At least a dozen were named, some with loss of life and others with all the crew rescued by nearby boats. There was even one sunk deliberately for the insurance.

During my fishing years, I had the use of my family's 18-foot Peterborough freight canoe with a little 3-horsepower outboard mounted on its transom stern. Sometimes my wife, Vicki, and I took it down the Campbell River from our home and across Discovery Passage to visit her family in Cape Mudge. It was a long and tedious trip. More fun was fighting our way up the river through the rapids. Using the skills that I had observed on the seine boat, I was able to take advantage of back eddies and, cautiously searching out the breaks in the current, to take that boat all the way up the river into the canyon. Except for possibly a jet boat, that canoe, with seine-boat skills, may still be the only power boat to have accomplished that.

When I returned to school in Vancouver that September of 1961, Vicki continued on the *DB No. 3* as cook until the end of the season. I had no idea about city life

Above: Alan in the Haig-Brown family's 18-foot Peterborough freight canoe. *Vicki Assu Robbins*

Left: A much younger Alan with his father, the writer and jurist Roderick Haig-Brown.

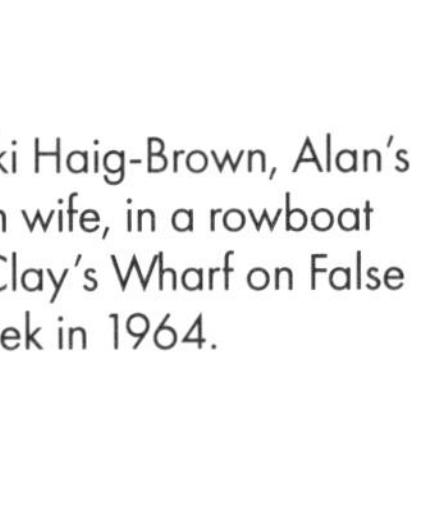

Vicki Haig-Brown, Alan's then wife, in a rowboat at Clay's Wharf on False Creek in 1964.

Living in False Creek near the National Harbours Board terminal introduced Alan to much of the coastal fleet, including these halibut boats. The *Zapora* was built in Prince Rupert in 1927 and, with new aluminum house, was still going in 2024. The *Island Queen No.1* was built by Wahl Boat Yard Ltd. in 1948.

or renting in Vancouver and could only think to walk down from my school to False Creek. There, I tried without success to get permission to live aboard a seine boat for the winter. I walked under the Granville Bridge to Granville Island, where I met the very nice son-in-law of the owner of a small marina called Clay's Wharf, who offered me rent-free accommodation in a little floathouse in return for caretaking the marina. The little house and office had been pulled up on the bank between the "Creek's" foul waters and a sawmill that slopped the coffee in your cup when they dropped another log on the carriage. Vicki and I lived three winters at Clay's Wharf. It was a great place for a fisherman to spend the winter. We had use of an array of boats, from a Briggs & Stratton–powered kicker to various row-boats and even a Sabot-class sailing dinghy. One winter, we lived aboard one of Clay's charter boats.

We spent a lot of time in the boats of the National Harbours Board fish boat docks next door. We met and made friends on a variety of trollers, long liners and seine boats. My water and fishing knowledge gave us entry into that community and taught me about other gear types and fisheries. But my proudest moment may well have been the night that Herb and Mitzi climbed aboard our floating home to have dinner.

Alan and his wife had the use of a "nice little kicker" when they wintered at Clay's Wharf in 1964. "We would run up to the Georgia Viaduct and walk to the Yen Lock in Chinatown for supper when we could afford it." *Vicki Assu Robbins*

CHAPTER TWELVE

Pleasing Side Trips

I think that one of the reasons that we were often bucking the tide when heading out for the usual 6:00 p.m. Sunday opening was that rounding up all the crew for a family boat is less than perfect science. Despite his superior knowledge of tides and fish, Herb lived by his own clock, and that clock was strongly influenced by family.

He loved catching salmon, particularly in his home waters of lower Johnstone Strait. This area is designated Area 13 by the Department of Fisheries. The northern section of Johnstone Strait, north from Kelsey Bay and up toward Cape Scott, is designated Area 12. Salmon of all species journey southward on their migration to their natal streams. In the 1960s, we routinely had three- and four-day fishing weeks.

It was quite common for the fishing to be good on Sunday night and Monday. By Tuesday afternoon, it would fall off. "I guess Hutch caught them all," Herb would laugh. I didn't know Hutch Hunt at the time, but when I met him in Fort Rupert years later, he reminded me of Herb. He had the same concern for family and the same love of fishing his home waters. But those waters, in Area 12 near the top end of Vancouver Island, tended to be more productive, and Hutch made a great deal of money. He used this to build beautiful modern seiners in both aluminum and fibreglass.

Meanwhile, back in Area 13, Herb delivered decent catches on Sunday and Monday nights. The Tuesday delivery to the company packer in Knox Bay, on West Thurlow Island, would be much smaller. I didn't fully understand it at the time, but the reduced catches were clearly a result of the interception of the fish farther north. At that point, an immigrant fisherman or his descendants, or an innovative settler like Byron Wright, for whom the whole coast was an invitation to discovery, might well have picked up his gear and travelled north to where the fish were. But that was not for Herb and most other Indigenous fishermen.

Before he bought his own boats, Herb Assu's brother-in-law Tony Roberts ran the big company boat *BC Pride*. In this photo, a deck hand is towing a manila line overboard to take out any kinks.

When the fishing fell off, we often went across the strait and into Blind Channel, between East and West Thurlow Islands. Sometimes in the fall we fished there for chum salmon. But farther into the channel was the site of an old salmon cannery operated by Quathiaski Canning. In 1935, it was sold to Japanese Canadian owners, who operated it as a saltery. This was confiscated by the Canadian government in 1942, and a couple of decades later nothing remained but the foundation. There was, however, a set of docks, a grocery store and a couple of houses. It was our treat to go up to the store and get a chocolate bar.

Out in the straits there was usually a cooling, summer westerly blowing, but in the sheltered channel, the summer heat, together with that of the engine, could make the fo'c'sle unbearable. I remember sleeping the night on the net pile while moored there, one of the perks of a table seine. Today Blind Channel operates as a resort for visiting yachts, but in those days it was a stopping point for passing loggers and fishermen and nice change from our usual anchorages.

I never asked Herb's reasoning for these "pleasure cruises," but I now imagine it as a celebration of his waters and his people's territory.

Other times we went on through Blind Channel and turned a short distance south to Shoal Bay or, more often, across Cordero Channel to Phillips Arm, on the mainland.

I was told that it was also the site of a no longer inhabited Kwakwaka'wakw village. We may have occasionally set the net there. I don't recall. More often two or three of us would take the skiff and go in search of an old mine shaft that was said to be not far back from the beach. There were also stories of the remains of a Japanese logging camp. Once, Mitzi came on deck as we were about to leave for the beach and cautioned, "You be careful and watch out for cougars in there."

I had grown up in the bush and once or twice seen a cougar that was running away. I was a bit surprised by her concern, but her people were of the water. The dark forests were the home of scary real and mythological creatures. As always, the visit ashore was fun and refreshing—though we didn't find the mine, a D^{z}unaq^{w}a wild woman, or even a cougar.

The greatest treat in Phillips Arm was found among the swaying fronds of eelgrass at low tide at the head of the arm. A river flowed in at one corner, and near there, at low water, you could row the skiff in the shallow over a fine sand bottom. Poking around among the eelgrass with a fish peugh, one could easily find big edible crabs. They scuttled around under the eelgrass and could be scooped into the skiff and taken back to the seiner, where Mitzi would have a big pot of water boiling on the galley's stove, and in due course we would sit around the deck and feast.

Chasing Phillips Arm crabs was not my only exposure to novel sea creatures. Once, three or four strange fish of 10 or 15 pounds came up in the net. I would recognize them now as some variety of tuna, probably albacore, a warm-water species rarely encountered in the notoriously cold waters of Johnstone Strait. At the time, neither I nor anyone else on the boat recognized them, and Herb ordered them to be thrown overboard. It was clear that these fish, in this place, were not a part of Herb's Indigenous knowledge.

There were plenty of fish—what we would call bycatch—that were familiar. More than once, when we were tempted by a good tide to set well before dawn, we were rewarded with hundreds of small rock cod tangled in our nets. And always there were jellyfish and the spiny dogfish. Another time a wolf eel nearly 6 feet long came up in the net and would gladly have sunk his fangs in anything that came within striking distance. Fishing herring in the shallow waters of Morse Basin, behind Prince Rupert, our heavy lead line dredged up dinner-plate-sized scallop shells. Birds had left turquoise abalone shells on rocky bluffs where we tied our net. Once, convinced by older crew members, I boiled a small squid for Herb to make up for a few of us going up to the bar in Steveston and making him wait. "He really likes squid," they said. Herb was not impressed when I took it to him on the end of a fork in the darkened wheelhouse. Another prank on the gullible *ṁaṁała* (white man).

Over the years, I had opportunities to learn more about Indigenous foods and the complexity of traditional trade that allowed for the sharing of them. One summer we were travelling south on the central coast. Herb pulled the boat into the dock at Klemtu, the Indigenous village on Swindle Island. He asked me to accompany him to a particular house in the village, where we obtained several recycled lard pails containing dried seaweed and several jars of soapberries.

We carried our treasures back to the boat and continued south. I was told by others that the people at Klemtu had access to a particularly good variety of seaweed, which they gathered and dried. Some of the crew on the winter herring had also purchased this or similar seaweed at Metlakatla, near Prince Rupert.

Some years later, I was working in the public school system and travelled often to Ulkatcho, on the western edge of the Chilcotin Plateau. There I learned of the Nuxalk people, who came up annually from Bella Coola with "gifts" of eulachon grease and seaweed. In return, their friends would take them hunting for moose and send them home with meat, processed buckskin and dried or jarred soapberries.

On visits to Bella Coola, I learned that in the 1960s and '70s it was common for local Indigenous basketball teams to travel by fish boat to the outer coast and villages at Bella Bella (Waglisla) and Klemtu. To trade for seaweed and other marine products, they would take with them eulachon grease from their river but also some of the precious soapberries that grow profusely on the high country of the Chilcotin Plateau. It would have been some of those berries that I purchased with Herb at Klemtu. It took me some time to realize that I was witnessing latter-day remnants of trading networks that once wove the Indigenous world together in a web of commerce so robust it established virtual trading highways, dubbed Grease Trails, that joined coastal groups to communities

Alan watches Herb Assu seal the lid on a can of sockeye using a hand-cranked canning machine aboard the *DB No. 3*. The cans were boiled in a huge pot on the galley stove. Once, it boiled dry and cans exploded, spraying fish all over the galley. *Vicki Assu Robbins*

in the BC interior. Early European explorers like Alexander Mackenzie learned to seek out and follow these well-trodden Indigenous thoroughfares.

Herb and Mitzi understood the channels and inlets of the coast in ways that no chart or coastal guide could. They knew them with a deep wisdom and a love that they quietly shared with family on the boat. If there weren't many salmon to be caught, there were other delights to be experienced. Once, Herb took the boat just a little farther north, to Ṫəqa (Jackson Bay). I had heard stories of this revered place from Autie, who had grown up in Salmon River. I spent many hours listening to Autie in her kitchen at Cape Mudge. One of the most powerful stories was of the Great Flood.

Ṫəqa was the place where the ancestors of many of the people now living at Cape Mudge and Campbell River had lived. We dropped the hook a short distance into the bay and climbed into the skiff. I think that this was the only time that I saw my skipper rowing the skiff. As we all did, he rowed standing up, facing forward and pushing rough-hewn 9-foot oars. The excitement on his face, I understand now, was of a deeply spiritual nature. We were returning to one of the home places of his ancestors. I still wish that I could have heard the stories that must have been playing in his head at that moment.

The shore bore little evidence of the long-vanished Kwakwaka'wakw village. In the intervening years, a white family had settled the natural meadows there and tried to farm. Like most such attempts, this one had failed for lack of markets, isolation from any support and being just too hard. The family had moved on, leaving a pleasant little house with a woodshed full of split stove-wood. Trees in an apple orchard had gnarled but fruit-laden branches.

More recently, a logging company had set up a small camp up at the head of the bay. Their yellow, rubber-tired skidders had churned up the meadow. There was no one around, and we didn't go near the equipment. I think Mitzi came ashore with us. I don't recall if there was any talk of the powerful story of the Great Flood, but this was not strange. Herb shared experiences, not stories. Or perhaps he and Mitzi recounted the story for each other in the Kʷak̓ʷala language. It told of a wise man who had survived the flood by anchoring his villages' canoes to a mountain.

I remember looking up at the tall, wooded peak at the mouth of the bay and imagining the rock that had anchored those canoes and now lay there. This was the rock to which the wise man had tied a long cedar-bark rope that he had made. He had been warned that a great flood would come. As his neighbours mocked him, he added length to this rope. When the flood came, he loaded his family into his canoe and tied it to the rope. His neighbours, no longer laughing, also boarded their canoes and hung onto the rope. They had a lot of food, but in the wind some of the canoes broke free and drifted north, while others drifted south. When the flood finally subsided, those canoes full of people had broken away, and became some of the north coast people and some of the south coast people.

I have never been back to Ṫəqa, but in the mid-1960s, at the time of our visit, Vicki and I were spending the winters living at Clay's Wharf in Vancouver's False Creek. It was in that Cold War time when we were assured that a nuclear war was a very real possibility. My fantasy plan was that, in the event of a nuclear attack, we would steal one of the charter boats moored there and seek refuge in Ṫəqa. Ṫəqa is that kind of a place. Maybe my idea also stemmed from the story of the man whose ancestors survived the Great Flood there. For Herb, it was a safe anchorage for his boat, but, I wondered, was it also a powerful cultural anchorage and refuge for his mind?

Top: Daniel Billy, a much-respected elder of the Cape Mudge people, seined herring on the *Ribac* in 1960.

Left: Daniel had the *Susan Laverne* built in 1975 by Cyril Thames, a noted builder of trollers.

CHAPTER THIRTEEN

Radar Rescue

When I started fishing in 1960, Herb and Mitzi had three sons; two more would be born later. Harvey and Wayne were only just teenagers and more interested in battling each other than absorbing their father's knowledge. Darryl, the youngest at the time, was three years old and delighted in working with twine and a sharp knife to assemble imaginary versions of the nets that he saw his father working with. I was always worried that he would cut himself or fall overboard trying to float a toy boat. But Herb and Mitzi, while keeping an eye on him, also allowed him to discover the bits and pieces of gear and knowledge that would add up to being a fisherman.

I think that it was the 1972 season that I was back on the boat, and Darryl, now a wiry 15-year-old, was my skiff man. We went north to fish pink salmon in Whale Channel. It was a lovely start to the summer. I had been teaching school in the Chilcotin the winter before, and at 31 years old the first weeks of the season were hard on my body. But even at that age, a body quickly adapts to the long days of pulling web and lines and nights of pitching fish. Darryl told me recently that his memory of

Opposite: A very young Darryl Assu on the *Departure Bay No. 3* at Cape Mudge in 1961.
Vicki Assu Robbins

Left: Darryl Assu setting the seine on the *Adriatic Star* in 2010.

that summer was of me singing Janis Joplin's "Oh Lord, Won't You Buy Me a Mercedes Benz." Ah, youthful exuberance.

There was also a favourite tie-up spot there. Someone had taken the time to cement a vertical 2-inch metal pipe into a crevice on the top of a big, rounded rock bluff. There were always big swells coming in off Caamaño Sound. The skiff man had to time his oar strokes to ride one of these up the sloping face of the rock. The tie-up man then jumped out before the falling wave swept the skiff back into the deep water. A few wraps on the pipe easily secured the beach end of the net. It was equally exciting getting out of there, but the skiff man had to be strong on the oars—a lot to ask of a young boy, but Darryl was fearless.

We spent one short weekend anchored out in Surf Inlet, just off Caamaño Sound. Herb's brother Ronnie and his wife, Lila, Mitzi's sister, tied up with us. The family was in its best place, at rest in a sheltered harbour far from modern-day Canada. My memory of that time is of sitting out on the deck smoking a cheap cigar in the motionless moonlit darkness framed by tall black forests, with only the clacking sound of cards being shuffled for a crib game on the galley table to disturb the silence.

The fishing years tend to merge one into the other, but the consistent memories are of the incredible warmth and security of that little 40-year-old wooden boat. Other weekends, when fishing the north, we went to Ocean Falls, Prince Rupert and, just once, all the way up Douglas Channel to Kitimat. Herb and Mitzi slept in the wheelhouse on bunks built across the back of the small space. Many wives worked on deck, running the hydraulic power block or performing other duties during a set, but Mitzi didn't. She kept her own cook's schedule. She got up after the crew, but by the time that we had the deck all cleaned and the gear ready for another set, Mitzi would have a massive breakfast waiting for us—and we would be ready for it!

Years later, while walking up Vancouver's Granville Street, I smelled something that triggered an intense rush of pleasure. It took me a moment to realize that it was the smell of bacon cooking in a restaurant blended with the diesel smoke of a passing truck. It placed me at work on the back deck of the *DB No. 3* with the smells of one of Mitzi's breakfasts melding with the boat's exhaust. Nothing like a scent to trigger a fond memory.

All marriages have their stress points, and living on a 57-foot boat with a half-dozen crew and often a small child must have created trying moments for Herb and Mitzi. But tensions were either rare or well concealed. I do recall one rainy day when I realized that Herb had not come down from his position up top for his occasional cup of coffee. Throughout the day I took him up coffee and snacks. He stood patiently with the rain dripping off the brim of a quality fedora that was not really designed for the coastal weather. I don't recall how long the tension lasted, and I have no idea how it was resolved, but the boat kept working and family life eventually returned to normal. With 13 kids in the family, there were often rotations, and on occasion Mitzi would send my young wife, Vicki, or one of her other daughters as cook.

I'm not sure what year we finally got radar on the boat. I had worked a little on coastal freighters and had become accustomed to radar as a near magical aid when travelling at night or in fog. It was early in the season, and we were again heading north to fish pinks. It was the middle of the night, and I was doing a four-hour wheel stint with a new white deckhand who we had hired off the dock in Campbell River.

We had passed Kelsey Bay in bright moonlight. At some point, I saw the light at Broken Islands off to our starboard, then a thick summer fog came down and lay heavily on the sea. I could still see the moon above the mast, and it reflected off the dew-drops on the radar's shiny white antenna housing beside me as I stood at the wheel. But the radar screen was in the wheelhouse below with our sleeping skipper and his wife. When you are steering blind in the fog, there is a tendency to turn away from the last land that you saw. I realized, by referencing the moon and the mast, that we were doing that now. I made the call to wake the skipper.

Herb came up top quickly, but he didn't turn on the unfamiliar radar. The fog had now obscured the moon, and we were totally enveloped. Herb navigated by a known shore just as his ancestors would have done in the days of the cedar canoes. Idling the boat, he headed for what he estimated somehow to be the Vancouver Island shore. It is all a steep drop from the mountain to the depths along this piece of coast, except for the mouth of Adam River, where a rocky shallow runs out some distance.

As luck would have it, that is where we came in. Just as we saw the trees, we heard the propeller hit rocks. Herb instantly reversed, but once we were clear of the shore and the engine clutch had been placed back in "forward," we felt a heavy vibration which had not been there before. We knew what this meant: the propeller had been bent when it hit the rock. We ran slow for the next few hours up to Alert Bay. We were able to get hauled out on the Alert Bay Marine Ways, the prop was fixed in short order, and we were ready to go again.

The shipyard crew had returned our boat to the water and tied it on the outside of a float. I had gone up on the bow to let the line go there. A couple of girls were playing in a little clinker-built rowboat on the inside of the float directly across from our bow. They were laughing and teasing each other by rocking the little boat. When the boat tipped and the girls fell into the water, I began to laugh also. One girl swam a single stroke and grabbed hold of the float. The other girl disappeared. I jumped onto the float and crossed it to look down into the clear waters. I could see the girl's long black hair spreading out a few feet below. I jumped in, reached down, and grabbed her. As soon as I had her up to the surface, there were several hands reaching down from the float. They lifted her out of the water. By the time that I climbed back onto the float, she was coughing up water and clearly recovering.

Our boat had already been untied, but Herb was holding it against the dock, and I was able to jump aboard. I was wet, pleased with myself, and being teased by the crew. Did the fog and the rocks at Adams River, which brought us to Alert Bay, lead to that girl's life being saved? I have often pondered on that since.

In the years before we got the radar, fog, especially combined with darkness, was a real threat. In the late 1950s, a halibut boat, the *Unimak*, had rolled over when it mistakenly passed between a tug and its tow. The tow line came up under the *Unimak* and caused her to flip. The story was that some of the crew trapped in the fo'c'sle could be heard tapping on the bottom of the hull for some time while the rescuers tried frantically to get the boat into the shallows. They were unsuccessful, and the men were lost. I have heard that story told in the fo'c'sle of our boat when travelling at night. Usually everyone decides to go up to the galley for another mug-up—safe from the danger of being trapped in the fo'c'sle.

Another time, again in the early 1960s, we were crossing the open waters of Queen Charlotte Sound. As darkness fell, a heavy fog set in a few miles north of us. Through the radio-phone's speaker up top, we could hear other boats, also without radar, checking back and forth with each other as they became surrounded by the fog.

One of our friends was chatting on the phone when suddenly we heard, "What is that? Back up! Back up!"

There was silence. Other boats called to ask what had happened. Finally, there came an answer, "We were travelling in the fog. Suddenly we saw the bow of a big barge approaching. We backed up, but the corner of the barge hit our bow. It tore some of it off, but we are not taking any water." With help and direction from some other boats, the damaged boat was able to run slowly in reverse until it got safely to Namu. I saw it there the next day with the bow sliced wide open. I was amazed that it had not sunk. I don't suppose that the tug even knew what had happened.

My favourite barge encounter came while travelling at night across the Gulf of Georgia between the Fraser River and Sabine Channel. Again, it was dark and foggy. We were a long way from shore, in the middle of the gulf, when suddenly I noticed the sweet smell of freshly cut cedar. I was perplexed until a huge barge laden with cedar logs emerged from the fog. We had plenty of distance between us so passed in the clear, and the smell of cedar in the gulf became just a pleasant memory.

Stories of being trapped in an overturned fo'c'sle could awaken fears, but for the most part the fo'c'sle was a warm and comfortable place of rest. When fishing hard, as we did for pinks up north, we mostly just tumbled into our bunks and fell right to sleep, knowing we had to drag aching bodies up a few short hours later.

But there were other times, anchored and waiting for the tide in Mitzi's Bay or travelling long distances in fair weather, when my bunk became a favourite reading place. My usual bunk was on the top and tight to the bow so that my feet could touch the boat's bow stem. If there was a good summer breeze blowing, we got a little circulation from a hatch that provided a second entrance or emergency exit from the fo'c'sle.

I always had a novel or two in my bunk, but I have found the fo'c'sles that I have been in are almost always equipped with the latest men's magazines. In the more staid 1960s, before the advent of *Penthouse* and *Hustler*, which leave nothing to the imagination, *Playboy* was the norm, with its airbrushed playmates and high-quality articles, exemplified by the famous *Playboy* interviews. Yes, I really did read them.

It was one of these interviews, a profile of John Wayne, that led to one of the few times I had an open disagreement with Herb. I knew that he was a fan of John Wayne, and one day he was going on about the actor. I had just read the now infamous 1971 *Playboy* interview in which Wayne was quoted as saying, "I don't feel we did wrong in taking this great country away from them [meaning Indigenous Americans] ... Our so-called stealing of this country from them was just a matter of survival. There were great numbers of people who needed new land, and the Indians were selfishly trying to keep it for themselves." I brought the magazine up top and read the quote to Herb. He looked genuinely disappointed and saddened, and I felt ashamed of my callow challenge to my skipper.

The Bee Hive Café, near the fishermen's wharf in Campbell River, had an excellent selection of magazines. When the new edition of the US-based monthly *National Fisherman* came out, I always bought a copy for the fo'c'sle. It didn't occur to me at the time, but reading that magazine on the boat led me to write and publish many articles in that and other fishing magazines from the 1980s until now.

The boat was our home, from magazines in the fo'c'sle to food in the galley. Regarding the latter, Herb kept an old Winchester Model 94 lever-action 30-30 in the boat's wheelhouse. Like the digging fork that was kept aboard for digging clams, the rifle was a tool for gathering food. Herb would shoot a little coast deer when the opportunity came along. This was usually on a shallow beach in the early morning at low tide, when the

Ripple Point, on Vancouver Island's lower straits across from Blind Channel, represented the southern limit of Herb Assu's home fishing waters.

deer come down to lick salt. The gun was decades old and well rusted from the salt air when I saw it. Someone said that Herb got it from the Canadian government when he was named a Coastal Guardian Watchmen during World War II. One day, near Ripple Point, he spotted a deer on a bluff high above the sea. He stood with his legs spread and his body, with its low centre of gravity, providing a rock-solid base. He fired a single shot.

The deer collapsed and began to tumble down the cliff until it dropped over a bluff and into the sea below. It was an incredible shot. "Take the skiff and go pick him up," he directed a couple of us. It was in my early years with Herb and, while I was impressed with the shot, I had reservations about the simple, efficient manner of the deer's death. It took me a little thought to understand that, while this was not the standard of sportsmanship that I had been raised with, it was the straightforward harvesting of food. It was no different than digging clams or picking huckleberries. It was simply a matter of getting food and one more way that the family fishing boat provided for the family. For me, it was also another important lesson in traditional Indigenous life on the BC coast.

I grew up on the Campbell River, in a house that my parents named "Above Tide" to signify that it was the first house up the river that even the highest tide did not reach. I thrilled to the salmon jumping all over the river in the fall. But this was nothing like the dorsal fin of a killer whale, or black fish, as the crew called them, rising alongside my skiff. More dramatic was when a whole family of killer whales cruised down along the beach then entered our pooling net before coming up on the other side, blowing triumphantly after having raided our haul and scared off all the fish they didn't catch. A thrilling show, but the price was a wasted set.

Wildlife sightings were commonplace but always elicited a nod or smile from even the most jaded crewman. Sometimes it would be schools of porpoise or dolphins surrounding the boat and surfing in our bow wave. As we drifted off Vancouver Island watching for salmon, we would see otters or mink scrambling along the rocky shore below the high-tide mark. The ubiquitous mink is a popular figure in Kwakwaka'wakw mythology, and Autie's tales of his many misadventures gave depth to my real-life experience, just as they would have to the people passing that way 100 years earlier by canoe.

CHAPTER FOURTEEN

Economics vs. Owner Operator

In the 1970s, a neo-liberal economic study of the BC commercial fishery concluded that there were "too many boats chasing too few fish." It laid out a process that eventually turned the ownership of the fishery over to a few fishing companies and processors. Today, with a few exceptions, the ownership of the commercial salmon and herring fishery is in the control of a single company. There have been many thousands of words written on the subject, but it was simple: boats, rather than people, would be licensed, and these licences would be transferable. In Alaska and on the Canadian east coast, there are "owner-operator" laws that require the licensee to be onboard the boat when fishing.

The value of the *DB No. 3*, now endowed with a valuable limited-entry licence, increased about eight times. At the same time, the roe herring fishery had been developed with Japan. The herring that I had fished for $8.80 per ton for fertilizer were now bringing $2,500 a ton for *kazunoko*, a Japanese delicacy made from herring roe, especially popular at New Year's. All this herring money was driving up the price of boats, no matter how old, as it was necessary to take a boat out of the fishery to build a new one. The *DB No. 3* was only 57 feet and narrow beamed, so it would pack about 30 tons, but licences under the new scheme were based on boat length, so a new fibreglass or aluminum seiner 57 feet long but much deeper and wider and capable of packing about 100 tons could be built using that same licence. In the 1980s, I was told by fishermen flush with money from the herring fishery, "My accountant said that I should build a new boat, which would be tax deductible, rather than give the money to the government."

For decades a corporate debt bondage had supported the fishery. In most cases, Indigenous fishermen were lent a small amount of money to buy and outfit a seiner. This loan was held by the company in return for having fish from their specific fishing

21908
ADR ATIC STAR

Above: The fibreglass *Pacific Pine* standing by while the aluminum *Western Wave* finishes up a set in Johnstone Strait in 2010.

Left: Before his death, Herb Assu purchased the *Adriatic Star*. It was big enough for herring, but when Herb died, the boat was seized. Darryl eventually bought it back, and the family still owned the boat in 2025.

Right: The *Phyllis Cormack*, March 1964. Herb Assu skippered this fine old double-decker on herring. Later, she sailed into history on the first-ever Greenpeace protest against nuclear bomb testing at Amchitka Island, Alaska. *Vicki Assu Robbins*

Below: The *Windward Isle* hauling back her gear near Camp Point in a tidy set, with no tide or wind to wrap the net around the bow or stern.

sites delivered to the company packers. The arrangement had colonial shades of the old Hudson's Bay Company, with Indigenous fur trappers owing their souls to the company store. In Herb's case, a good part of his value to the company was the skill and knowledge to make that set at the Slide and along his familiar stretch of coast.

The companies had long financed aggressive immigrant fishermen—for whom the whole coast was a potential fishing ground—to buy ever bigger and more modern boats. This was not so much a plan as an inevitable result of settler-owned fishing companies, matching fishermen to boats that suited their fishing culture. The addition of nylon gear, more powerful engines, and net drums reduced the value of the local knowledge possessed by many Indigenous fishermen. At the same time, the costs of limited-entry licences and technological innovations loaded greater debt on the fishery.

My last season seining salmon with Herb was in 1973. I had already decided that I didn't want to spend another winter fishing herring. Somewhere I had heard of a private institution in Vancouver, on West Tenth near Granville Street. It was called Shurpass Pacific College and specialized in helping dropouts patch up holes in their education. It sounded like what I needed to finish my high school and get off the boats. Seining had been a wonderful introduction to work and life, but after ten years I was ready to test my shore legs.

Herb carried on in the winter herring fishery, successfully running other people's large herring seiners, like the *Phyllis Cormack*, which would become famous as the boat Greenpeace used to protest nuclear testing at Amchitka Island, Alaska. Eventually, he sold the *DB No. 3* and its licence for a good bit of money and bought the big, ironically named *Adriatic Star*. The debt load was many times that of the boat that I started fishing on in the 1960s.

Consolidation had been taking place elsewhere in the fishery, with the Norwegian-Canadian-owned Nelson Bros Fisheries Ltd. being sold to BC Packers Ltd. This latter company was the result of many smaller companies being bought up over the decades before finally being purchased by the Weston family, an eastern Canadian food conglomerate. Where Nelson Brothers was built and owned by men who had fished and knew the business, the owner of BC Packers was an eastern blueblood noted for his polo playing with English gentry. A long way, geographically and spiritually, from Johnstone Strait and the Slide. Loans were no longer held by fishing companies that understood the fishermen, they were transferred to banks.

When Herb died, in 1983, Mitzi was not able to pay their significant bank debt. The family lost the house in Richmond that they had bought years before, and they lost the *Adriatic Star*. After a long and protracted struggle, Darryl, who had learned to make the set at the Slide and skippered other people's boats, was able to buy back the *Adriatic Star*. But the fishery of the 1990s and later bore little resemblance to the era of Herb and his ancestors.

Further consolidation had led to the Canadian Fishing Company, a Jim Pattison company, buying the fleet and licences of BC Packers. The Weston family kept much of the land where canneries, shipyards and workers' homes had stood. Where feasible, this was subdivided and built over with luxury homes—one more round of corporate profit

SPLENDOUR
22029

Above: The wooden Wahl-built seiner *Scanner* about to be demolished at Port Hardy in 2013. The aluminum and the 6-71 diesel engine would be salvages, but the wood would go to the landfill—the fate of those that don't burn or sink.

Left: The *Splendour* at anchor at the Slide in 2001, waiting for a chum salmon opening with Mark and Darryl Assu. The *Splendour* was built in Ladner, BC, by Mario Tarabochia in 1940 for the Martinolich family.

Opposite top left: As environmental depredation and other factors threatened some salmon runs, the sorting box was introduced so that non-targeted species such as chinook or steelhead could be released. Sometimes they were held in the water of a recovery box for a short time.

Opposite top right: The late Judy Assu Wagner preparing a meal on the seine boat *Splendour*.

Opposite bottom: The traditional wooden *Taplow* and the newer aluminum *Summer Hawk* show the dramatic increase in packing capacity in similar lengths.

taken from the BC fishery. At the same time, one person now owned the great majority of the salmon and herring licences. The Canadian Fishing Company purchased Darryl's debt from BC Packers and told him that he would thereafter be fishing for them. Darryl refused, and they demanded immediate payment of several hundred thousand dollars. Darryl had no choice but to sell the family salmon licence to pay the debt.

As Darryl told me this account, I was reminded of the fact that Canadian resource management loves a monopoly: first the Hudson's Bay Company, whose Royal Charter gave them exclusive trading rights to all the waters that drain into Hudson Bay; then the Canadian Pacific Railway was granted massive lands along their railway to BC. In the first half of the twentieth century, many independent logging families gained pieces of timbered lands and logged independently. This largely ended with the ill-named Tree Farm Licence (TFL) system, which left most small independent companies competing for work and dependent on contracting for giant TFL-holding forest products corporations.

By the 1960s, for many coastal families, the only independent resource industry remaining was fishing. Even with the weight of company debt, there were always truly independent fishermen. The corporations, working with friendly Fisheries ministers, like Jack Davis, were finally able to break that independence with the limited-entry and transferable boat licences. In western Canada, unlike Alaska, there was no requirement that the owner be on the boat. Technology made the fishery less dependent on local and heritage skills. In the 2010 fishery, Canadian Fishing Company stacked multiple licences on a single company-owned boat and dictated a share system that favoured their profit. As with the fur trade, the railway grants, and the tree farm licences, the granting of fish quotas was based on land and resources that had been—and arguably still were—the property of Indigenous nations.

By this time, I was writing about commercial fishing for various trade journals. The fishery that I had worked in was no longer recognizable. I went out fishing with Darryl on several different boats to research and write stories. Once, in the early 2000s, I went on a classic boat named *Splendour*, built in 1940 by Mario Tarabochia of Ladner. Darryl and his brother Mark were sharing command, with their sister Judy doing the cooking and a couple of relatives for crew. We went out early on the Sunday and spent a good part of the day at anchor just off the Slide. We even took the skiff and went ashore. As in the old days, there were lots of fish showing throughout the afternoon, but fewer by the 6:00 a.m. opening. As their father and grandfathers had done before, the brothers set promptly at 6:00. It was a good tide, and they held the net open until a glorious shout came across the water from the beach man, "Jumpers! Going in!"

It was a splendid replay of the historic set. And yet it was diminished by the short one-day opening and the small number of fish being caught. To protect the small run, the Department of Fisheries shut down the fishery to allow enough escapement for the in-river Indigenous fishery and the requirements of the spawning grounds to assure the continuity of the ancient runs. The fishery was curtailed to protect the salmon, but the other threats, from logging to urban development and rural mining, continued seven days a week with business as usual.

SEL

SUMMER HAWK
SUMMER HAWK
TAPLOW
29285

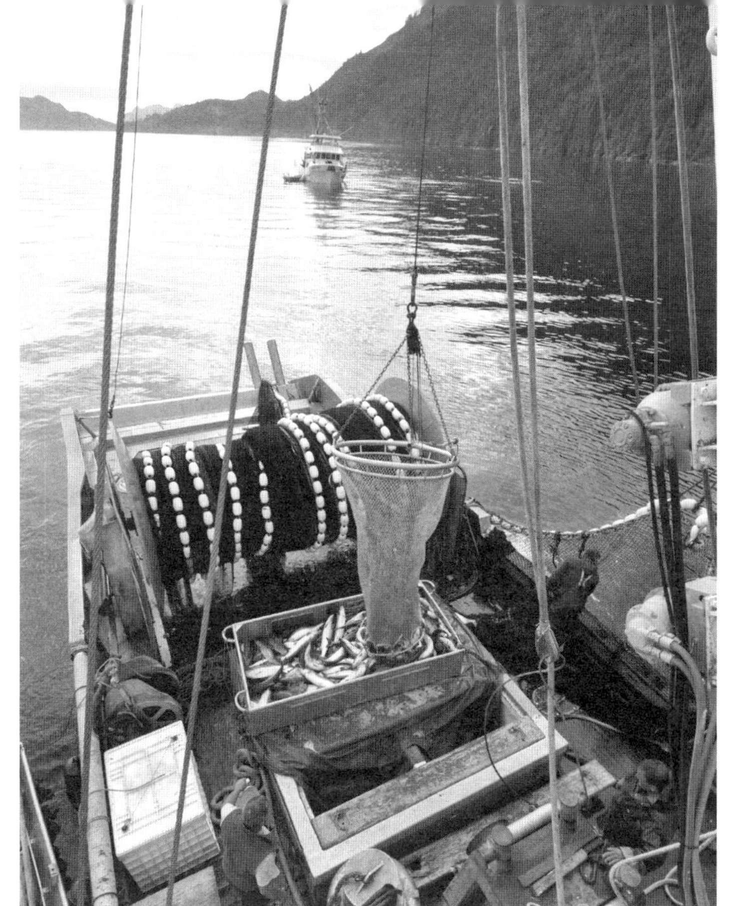

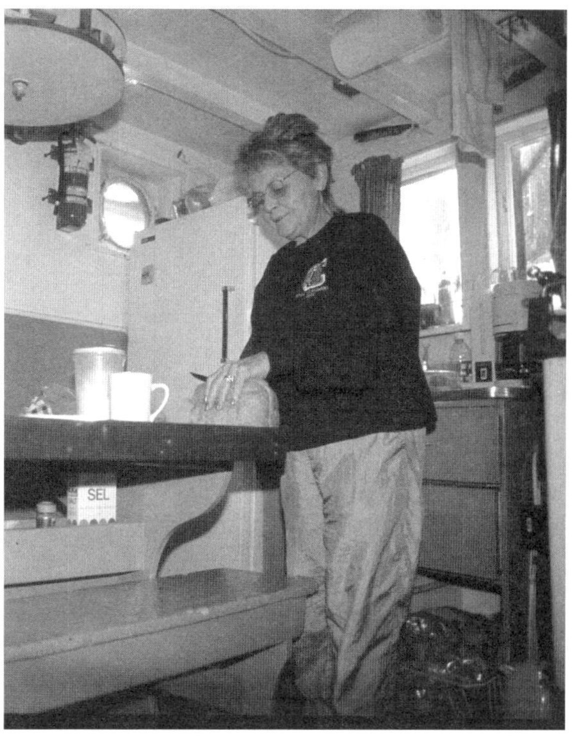

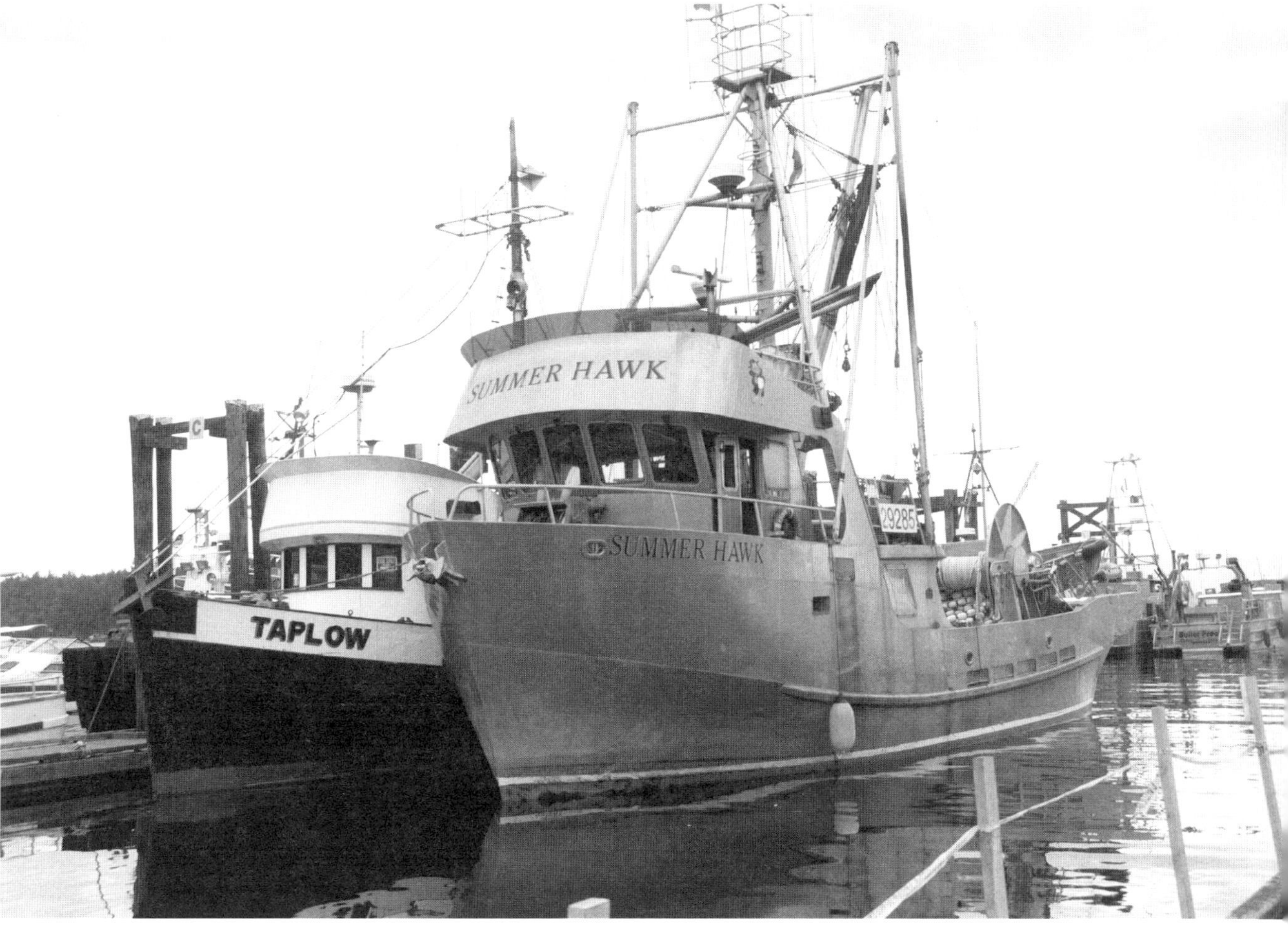

CHAPTER FIFTEEN

One More Time

In 2010, Darryl invited me to go out with him again with a crew he had pulled together for the trip. It was an entirely different era. A miraculously huge—not to be repeated to date—run of Adams River sockeye had shown up. The fishing company and the government had cooperated to declare a quota of 3,000 sockeye per licence, per trip. Darryl had gained access to a single salmon licence, and we headed out to fill the quota.

Not uncommonly, the tides and some storm-class winds were kicking up a mess in the seas off Cape Mudge. Darryl decided to shelter around Hornby Island for a few hours. Eventually we pulled up the anchor and headed north. One of Darryl's sisters, along as the cook, came out of the accommodation area and, seeing the bench in the galley overturned and three or four of us sitting dejectedly at the galley table while storm-driven white caps swept by the windows, said, "I didn't sign up for this. Listen to the damn timbers creaking!" To which one of the crew replied, "You don't have to worry until they stop creaking."

As previously noted, the waters south of Cape Mudge rank up there with Seymour Narrows as challenges for fish boats and tugs. It is a place where wind and tide can turn a pleasant summer day into a maelstrom of terror. The tides that sweep in from the north and south ends of Vancouver Island meet to cause confused seas with enormous standing waves. A Seattle tug skipper who had spent time in the deadly Bering Sea crab fishery told me, as we approached the Cape in a modern 100-foot tug, that he has been as scared off Cape Mudge as he ever was in the Bering.

Many years before this passage with Darryl, when he was still a young teenager, we had a bad passage with his dad as skipper. We were returning with the boat from the Fraser River. It was blowing a nasty southeaster. Herb had decided to take the more cautious and protected route up Malaspina Strait between the east side of Texada Island and the mainland. All went well until we had to make that perilous passage from around Powell River across the south end of Quadra Island. My memory of the crossing is limited. At some point I went in the wheelhouse to see Herb's bulk wedged into a corner with his eyes straight ahead and a hand on one of the wheel's wooden spokes, steadying the boat's motion. He said to check for water in the hold and to start pumping with the deck pump if required. Darryl and I removed one of the hatch covers and saw the water that Herb could feel without looking.

The BCP50, skippered by Darryl Assu in the late 1980s.

The waves were hitting us broadside and washing over the deck, so I stood on the port bulwarks with a good hold on the rigging and began working the big iron pump handle located just behind the cabin. Things were not looking good. We didn't have that far to go, but it seemed to take forever. I recalled all the stories, like the loss of Buddy Roberts, Vicki's uncle, and wondered if this was how it would end. I could see the lights of the Rockland subdivision on the hills just south of Campbell River. Vicki and I had recently spent a pleasant evening in the living room of friends who lived there. The contrast between that secure living room and my tenuous perch on the bulwarks, with wind and tide battering the boat, remains indelible in my mind. We made it home of course.

I often say that my years of fishing were more important to my learning than my time at university and, as it happens, they opened up my career for the latter half of my life. In 1986, my family and I moved to Vancouver. I had published a couple of articles about BC fish boats that got me the job as editor of a new magazine called *West Coast Fisherman*. That led to other work that has taken me on board boats from Borneo's Rajang River to the Amazon, China, Europe and the Gulf of Mexico.

Alan in his magazine photographer days.

In 1967, Vicki and I completed our first-year teaching school on Quadra Island. We bought a truck and drove to Expo 67 in Montreal. On the way home, I had thoughts of working on a prairie harvest, but once our pickup was headed west there was no stopping. As soon as we got back to Campbell River, we went down to the docks to see the fleet off to their Sunday opening. Herb backed the *DB No. 3* a few feet off the float. He looked down at me and said, "Go home and get your slicker gear, we'll wait for you."

I was back out on the boat for salmon most years up until 1973. Herb had passed on by the time that I started writing about fishing, but Darryl took me out a number of times on a variety of different boats that he was running. There was once, on a big double-decker named *BCP50*, that we went just off the Fraser River for a rare seine opening in the Gulf of Georgia. I was standing up top taking pictures of the crew getting ready to brail a good set.

Suddenly a Fisheries patrol boat came alongside and a well-known Fisheries officer boarded our boat. She walked past all the crew working on deck and came to me on the cabin top.

"Do you have a fishing licence?" she asked.

"No, I'm just taking pictures for a magazine article."

"You need a licence or a permit to be on board an active fishing vessel," she declared triumphantly and wrote me out a ticket that required me to attend court in Steveston some weeks later.

I was told by Darryl that I had just met the notorious "Sockeye Sue." It was a short opening, and we headed back to Steveston later that afternoon. As soon as we had tied up, three of the crew got off and headed uptown. Darryl explained that they had gone up to the Department of Fisheries office. He had just hired them, and none of them had a licence!

A month or so later, I attended the court on my scheduled date. I was waiting my turn when fisherman-lawyer-poet John Skapski came into the court and asked me what I was doing there. I told him my tale of woe. He had the judge dismiss the case. I left with one more fishing story. John's great book on his own life in fishing, *Green Water Blues*, has a special place on my shelf.

Darryl Assu on the brailer handle on the *Splendour* in the late 1990s. The principles of seining remain the same, but the technology has changed a lot. The chum salmon are brailed into a sorting box so that non-targeted chinook and steelhead can be released.

Brailing sockeye on the *Adriatic Star* in Johnstone Strait, 1990.

Meanwhile, on the 2010 trip with Darryl, there were lots of fish showing in the usual places in the straits, but it took a few sets to get our quota. As always, it was a thrill, but now it was overlaid with a sadness. The quota system, the temporary crew, the quick trip—and again the quota—all conspired to make it so very different from the old family-boat days with Herb. Except there were moments that brought those times back. When I was up top with Darryl, he nosed the big boat in along the steep rock beach, and I watched him watch the tide. His face bore the same calm, thoughtful countenance that I had seen on his father's face so often. There is magic afoot when the person, the sea, the fish and the tides find harmony.

By 2024, the aging *Adriatic Star*, now under the command of Darryl's younger brother Mark, was still scraping by with some contracts to pack salmon or herring. Wooden boats have become an anachronism in the fishery, just as the importance of the set at the Slide has been diminished by technology and reduced runs. The old boats

The lovely *Istra* in Johnstone Strait in 2010. She was hauled ashore and crushed in 2024.

Left: *Cape Flattery*, built in 1989 by Shore Boat Builders Ltd. in Richmond, BC, was owned by BC industrialist Jim Pattison. When Pattison became active in Alaska, he was forced to sell many of his boats to local fishermen because under Alaska's owner-operator law, a boat must be owned by the person who fishes it. Many wish BC had a similar law.

Below: Built by Matsumoto Shipyards Ltd. in 1974, the *Santa Cruz* waits her turn to set. Mark Assu was skipper in 2010.

are virtually impossible to insure. With fear of fuel leaks should they sink and incur expensive clean up, many owners are having them taken ashore and demolished.

Much of the BC seine fleet has been sold to owners in the US, where the licences to fish are carried by the fisherman and not the boat. Some of these boats have been sold to Alaskan fishermen who fish for the Pattison-owned Alaska General Seafoods. By and large, the Alaskan fishery remains healthy, while the much-diminished BC fishery continues to struggle. In the spring of 2024, two fine wooden seiners, the *Istra* and the *Sea Master*, were hauled out of the water, crushed by a backhoe and loaded into dumpsters. Apparently, the licences had been bought back by the government and turned over to First Nations. The government paid to destroy these masterworks of the shipwright's art, fashioned from old-growth BC timber the like of which will never be seen again.

Dinner is served on the *Adriatic Star*, 2010. Left to right: Jay Grabher, Denise Assu, Tarzan Scow, Sean Wagner, Al Wagner and Darryl Assu.

CHAPTER SIXTEEN

Mitzi and Herb

Mitzi outlived Herb by decades. She was my mother-in-law, but she was so much more over the 60 years that I knew her. It was in the winter of 1959 that Vicki, her second daughter, took me home for supper at their house in Cape Mudge. It was my first time visiting that Quadra Island village. We had a fine meal and then Mitzi laughingly said, "We will let the *m̓am̓ała* do the dishes."

It was my introduction to the cross-cultural humour that would accept me but mark me as "white" in the Indigenous community. By the next summer, Vicki and I were married, and I went to work on the *San Jose*, the seine boat that had belonged to the late Dan Assu and was that year skippered by Herb.

When you spend months on a small boat with a half-dozen or so other people, you get along or get off. Everyone has their job, and they do it or leave. Mitzi was the cook. When I started on the boat, my brother-in-law Darryl was only four years old. It seems there was always a little one or a reluctant teenager. Mitzi did all of those mothering things in a tiny space, with endless patience and laughter, with only an occasional word of frustration in Kʷak̓ʷala: "*madᶻos!*" ("Oh for God's sakes!")

Mitzi Assu was my shipmate and so much more—from Cape Mudge to the house in Richmond that she lost after Herb died, and then her house at Quinsam. We both suffered together through the loss of loved ones. In my youth, I knew and learned from Mitzi's mother and her grandmother. To tell all about this powerful and lovely woman would be to tell much of my life: she was a second mother to me.

When I talk with other fishermen my age and even a bit younger, one of us will usually conclude the visit with, "We are so damned lucky to have lived when we lived."

Someone will add, "And you were fortunate to have fished with Herb Assu. He was a true gentleman."

Herb, who was born in 1921, died of diabetes in November of 1983 after a long period of decline. He was indeed a gentleman, but he was more. He was my professor of the waters and of life. For some of my commercial fishing years, I also attended

university. The learning on the boat was a slow, immersive process, while university was an intense rush of lectures and undigested information. Like a slow-cooked roast, the boat learning was rich in flavour and nutrition. The university, with its diverse lectures and specialized faculties, was more like a cafeteria meal served on a segmented plastic plate.

In the ensuing decades of life, the fishing education has served me well as a public-school teacher and as a writer. It has allowed me to navigate complex cross-cultural understandings worldwide, and it has guided me in virtually every aspect of my adult life.

In my fisherman years, the practical pull of the tides and the push of storms always merged with the mythological stories of the old people, especially Autie's. Sitting at her kitchen table overlooking Discovery Passage, she told me stories that my mind set in the waters that Herb taught me to feel. Her stories of ƛisəlagiƛ̓akʷ, the mink, deepened my appreciation of the actual creature that scrambled mischievously along the shore where I tied my beach line. ƛisəlagiƛ̓akʷ set out to marry. First, he married the bull kelp, but when the tide came up, he nearly drowned, so he married the barnacle, but when he beat her for her stoic silence, he only cut his own hand. There was a caution in many of the stories that only as an old man have I begun to appreciate for their depth in our complex world of currents and upwellings. The surface doesn't always reveal the depth. Herb's calm patience and quiet understanding provided me with an example whose depth of meaning has only grown with the passing years.

Opposite: Alan visiting Mitzi Assu at her home on the Quinsam reserve near Campbell River in 2018. *Celia Haig-Brown*

Following pages: Built by W.R. Menchions & Co. Ltd. in 1926, the *W No. 8* will fish her 100th year in 2026.

Seiners and Alaskan cruise ships, such as the *Celebrity Millenium* pictured here, keep wary company in crowded Johnstone Strait.

24124
W-8

Index

Note: **Bold** page numbers indicate a photo on the corresponding page.